T0185641

Application of Visible Light Wireless Communication in Underground Mine

Simona Mirela Riurean • Monica Leba
Andreea Cristina Ionica

Application of Visible Light Wireless Communication in Underground Mine

 Springer

Simona Mirela Riurean
Universitatea din Petrosani
Petroşani, Romania

Andreea Cristina Ionica
Universitatea din Petrosani
Petroşani, Romania

Monica Leba
Universitatea din Petrosani
Petroşani, Romania

ISBN 978-3-030-61410-2   ISBN 978-3-030-61408-9   (eBook)
https://doi.org/10.1007/978-3-030-61408-9

© Springer Nature Switzerland AG 2021
This work is subject to copyright. All rights are reserved by the Publisher, whether the whole or part of the material is concerned, specifically the rights of translation, reprinting, reuse of illustrations, recitation, broadcasting, reproduction on microfilms or in any other physical way, and transmission or information storage and retrieval, electronic adaptation, computer software, or by similar or dissimilar methodology now known or hereafter developed.
The use of general descriptive names, registered names, trademarks, service marks, etc. in this publication does not imply, even in the absence of a specific statement, that such names are exempt from the relevant protective laws and regulations and therefore free for general use.
The publisher, the authors, and the editors are safe to assume that the advice and information in this book are believed to be true and accurate at the date of publication. Neither the publisher nor the authors or the editors give a warranty, expressed or implied, with respect to the material contained herein or for any errors or omissions that may have been made. The publisher remains neutral with regard to jurisdictional claims in published maps and institutional affiliations.

This Springer imprint is published by the registered company Springer Nature Switzerland AG
The registered company address is: Gewerbestrasse 11, 6330 Cham, Switzerland

Preface

Bill Gates, said, "We're changing the world with technology." Due to the present continuous unexpected progress of science, we are witnessing amazing technological discoveries with huge impact on us and on our daily activities. We move faster than ever, we are eager for high-speed, remote communication, and we are most of the time connected by Internet.

We are not far from the moment when remote, human-to-human (H2H) communication has been a novelty. Today, human-to-machine (H2M) or even more advanced, machine-to-machine (M2M) communication has become reality. Smart cities (SCs), smart houses (SHs), smart driving, smart personal devices, and so on, rely on local or remote wireless data communication based on electromagnetic (EM) spectrum. The Internet of Things (IoT), an already worldwide reality, needs high-speed data transmission both indoors and outdoors. All smart devices, that we use every day, need fast and secure data traffic.

As Cisco Visual Networking Index data traffic forecasts, by 2022, there will be more than 12.3 billion mobile devices connected, exceeding the foreseen population (of 8 billion) by 50%. Therefore, the wireless transmission based on radio frequency (RF) is becoming more crowded every day and the RF spectrum crunch will soon become reality. In order to relieve the limited and expensive RF spectrum, there have been explored lately new and equivalent wireless communication technologies.

Many alternative solutions have been tried to support this exponential growth and "hunger for Terra bytes" of wireless data communication. One of the most reliable and favourable emerging alternatives that offers many advantages over RF communication is the optical wireless communication (OWC) technology. Since the most data traffic takes place indoors, OWC proves to be a suitable and reliable partner for the today's classical wireless data communication based on RF (such as Wi-Fi, cellular, Bluetooth, BLE, WiMAX, ZigBee, ZWave, and so on). OWC covers Infrared (IR), visible light, and ultraviolet (UV) wireless data transmission.

Year 1976 has been an important one for IR transmission, since *Gfeller and Bapst* demonstrated for the first time, fast, accurate, and secure wireless optical communication at up to 125Kbps data rate.

More than two decades later, in 2003, *Tanaka et al.* validated years of theoretical research and developed a VLC system that conveyed data over illumination at a data rate of 400 Mbps.

Since then, researchers worldwide presented important works in VLC area. From a 500 Mbps data rate in a Line of Sight (LoS) topology demonstrated in 2006 to 15.73 Gbps validated in a research laboratory in 2019, step by step, totally or partially, the many drawbacks of the VLC systems have been overcome.

Numerous VLC projects with remarkable ideas have made possible various implementations in different fields. A number of areas, such as those where wireless data communication based on RF is limited (airplanes or health facilities) or forbidden (nuclear power plants, petrochemical industry, or coal underground mines with high explosion risk) as well as other overcrowded places with high demanding wireless connectivity, are appropriate places for VLC to be used.

For some years, efforts have been made to deploy systems based on VLC, in different fields, as indoor data wireless communication or indoor positioning systems and navigation (e.g. in museums, supermarkets, or underground). VLC is also envisioned to become an advanced technology embedded in smart medical devices, home appliances, or various other devices. Vehicle-to-vehicle (V2V), infrastructure-to-vehicle (I2V), or vehicle-to-infrastructure (V2I) types of communication, as well as underwater and underground VLC are promising research areas, as well.

The miner's profession is one of the most dangerous one worldwide, mainly because of the specific workplace characteristics. For the underground mining area activity, no matter the type of exploitation, ore, coal, salt, or any other kind of minerals, a significant volume of research has been dedicated to identifying the main risk factors for health and workers' safety and decreasing, as much as possible, their negative effects.

In order to improve workers' safety underground, many different technologies have been already theoretically proposed and some practically applied. The workers' position underground and continuous monitoring of risk factors in real time are important issues in underground operation, where the conditions are continuously changing.

More and more internationally well-known companies run projects aiming to add to the core illumination function, data communication, and to solve challenges for high-speed, highly secure wireless connectivity.

The VLC technology promises to provide, in the near future, a safer, faster, and greener underground data wireless communication system. As fast this technology develops, within a few years, we expect to see VLC together with other wireless complementary technologies creating a new ubiquitous computing platform. Under this forthcoming integration, every device large enough to include an LED, a transmitter driver, and a light sensor can be connected and powered by VLC.

Petroşani, Romania   Simona Mirela Riurean
Petroşani, Romania   Monica Leba
Petroşani, Romania   Andreea Cristina Ionica

Abbreviations

ACO-OFDM	Asymmetrically Clipped Optical OFDM
ADC	Analogue-to-Digital Converter
AoA	Angle-of-Arrival
AP	Access Point
APD	Avalanche Photo Diode
AR/VR	Augmented Reality/Virtual Reality
ARPANET	Advanced Research Projects Agency Network
AS	Accelerometer Sensor
ASCII	American Standard Code for Information Interchange
ASIC	Application-Specific Integrated Circuit
ASK	Amplitude Shift Keying
ATA	Analogue Telephone Adapter
AWG	Arbitrary Waveform Generator
AWGN	Additive White Gaussian Noise
AZO	Aluminium Doped Zinc Oxide LED
BAN	Body Area Network
BER	Bit Error Ratio
BJT	Bipolar Junction Transistors
BLE	Bluetooth Low Energy
BRDF	Bidirectional Reflectance Distribution Function
BW	Bandwidth
CAP `	Carrierless Amplitude and Phase
CB	Coherence Bandwidth
CCR	Constant Current Reduction
CD	Chromatic Dispersion
CIR	Channel Impulse Response
CM	Coded Modulation
CMOS	Complementary Metal-Oxide Semiconductor
CRD	Colour Rendering Index
CSK	Color Shift Keying
CSMA/CA	Carrier Sense Multiple Access/Collision Avoidance

CU	Ceiling Unit
DAC	Digital-to-Analogue Converter
DC	Data Centre
DCO	Direct Current Biased Optical
DCO-OFDM	DC biased Optical–Orthogonal Frequency Division Multiplexing
DD	Direct Detection
DFE	Decision Feedback Equalizer
DFT-s-OFDM	Discrete Fourier Transformation spread OFDM
DMT	Discrete Multitone
DS	Delay Spread
DSP	Digital Signal Processor
DSP	Digital Signal Processing
DU	Desktop Unit
E/O	Electrical/Optical
ED	Excess Delay
EMI	Electromagnetic Interference
ENOB	Effective Number of Bits
EPSRC	Engineering and Physical Sciences Research Council
FDM	Frequency-Division Multiplexing
FEC	Forward Error Correction
FFT	Fast Fourier Transform
FIR	Finite Impulse Response
FL	Fluorescent Lamps
FMF	Few-Mode Fibre
FoV	Field of View
FSO	Free Space Optics
FWHM	Full Width at Half Maximum
FWHT	Fast Walsh–Hadamard Transform
GaN	Gallium Nitride
GBDM	Geometry-Based Deterministic Models
GBSM	Geometry-Based Stochastic Model
HCM	Hadamard Coded Modulation
HL	Halogen Lamps
HPC	High-Performance Computing Infrastructure
IB	Incandescent Bulbs
ICI	Inter-Carrier Interference
IEEE	Institute of Electrical and Electronics Engineers
IETF	Internet Engineering Task Force
IFFT	Inverse Fast Fourier Transform
IM	Intensity Modulation
IM/DD	Intensity Modulation/Direct Detection
IoT	Internet of Things
IPS	Indoor Positioning Systems
IR	Infrared

IS	Image Sensor
ISI	Intersymbol Interference
JEITA	Japan Electronics and Information Technology Industries Association
LED	Light Emitting Diode
LFS	Leaky Feeder System
LiFi	Light Fidelity
LoS	Line of Sight
LTE	Long-Term Evolution
M2M	Machine-to-Machine Communication
MAC	Medium Access Control
MCBM	Modified Ceiling-Bounce Model
MCMT	Multi-Carrier Modulation Techniques
MIMO	Multiple-Input, Multiple-Output
MMCA	Modified Monte Carlo Algorithm
MOSFET	Metal Oxide Semiconductor Field-Effect Transistors
M-PAM	Multi level Pulse Amplitude Modulation
M-PPM	Multi level Pulse Position Modulation
NFC	Near-Field Communication
NFT	Nonlinear Fourier Transform
NLOS	Non-Line of Sight
NLSE	Nonlinear Schrödinger Equation
NON GBSM	Non Geometry-Based Stochastic Model
NRZ	Not Return to Zero
N-SC	Nyquist Single Carrier
OCC	Optical Camera Communication
OFDM	Orthogonal Frequency-Division Multiplexing
OLED	Organic LED
OOK	On/Off Keying
oRx	optical Receiver
oTx	optical Transmitter
OWC	Optical Wireless Communication
PAM	Pulse Amplitude Modulation
PAN	Personal Area Network
PAPR	Peak to Average Power Ratio
PCB	Printed Circuit Board
PCM	Pulse Code Modulation
PD	Photodetector
PDM	Polarization-Division Multiplexing
PHY	Physical layer
PIN PD	Positive Intrinsic Negative Photodiode
PLC	Power Line Communication
PLED	Polymer LED
PMD	Polarization-Mode Dispersion

PoE	Power over Ethernet
PON	Passive Optical Network
PPM	Pulse Position Modulation
PS	Phase-Shifted
PSD	Power Spectral Density
PSK	Phase Shift Keying
QAM	Quadrature Amplitude Modulation
R&D	Research & Development
RF	Radio-Frequency
RFID	Radio-Frequency Identification
RGB	Red Green Blue
RMS DS	Root-Mean-Square Delay Spread
RMS	Root-Mean-Square
RS	Reed–Solomon
RS GBSM	Regular-Shaped GBSM
RSS	Received Signal Strengths
SC	Smart City
SCMT	Single-Carrier Modulation Techniques
SDM	Space-Division Multiplexing
SH	Smart Houses
SLD	Super Luminescent Diode
SNR	Signal-to-Noise Ratio
SPAD	Single Photon Avalanche Detector
SR	Spectral Response
SSL	Solid-State Lighting
TCP/IP	Transmission Control Protocol/Internet Protocol
TDM	Time-Division Multiplexing
TDoA	Time-Difference-of-Arrival
TIA	Trans-Impedance Amplifier
ToA	Time of Arrival
TTE	Through-the-Earth Transmission
UHF	Ultra High Frequency
UOWC	Underwater Optical Wireless Communications
UP&MS	Underground Positioning & Monitoring System
USBM	US Bureau of Mines
UTP	Unshielded Twisted Pairs
UV	Ultraviolet
UWB	Ultra-Wideband
VCSEL	Vertical-Cavity Surface Emitting Laser
VHF	Very High Frequency
VLC	Visible Light Communication
VLCC	Visible Light Communication Consortium
VLD	Violet Laser Diode
VoD	Ventilation on Demand

VoIP	Voice over Internet Protocol
VPPM	Variable PPM
WDM	Wavelength Division Multiplexing
WiMAX	World Interoperability for Microwave Access
WLAN	Wireless Local Area Networks
WLED	White LED

Contents

List of Figures

List of Tables

Chapter 1
OWC Developments and Worldwide Implementations

1.1 Short History of OWC Concepts and Terms

Optical wireless communication (OWC) refers to any kind of communication based on *light* as a wireless transmission medium.

Visible light communication (VLC), free space optics (FSO), optical camera communication (OCC), light fidelity (Li-Fi) and infrared (IR) are all different parts or applications of OWC.

VLC refers to an artificial light source, a light emission diode (LED) (picocells, attocells or laser diodes—LD) that uses illumination to send data piggybacked by the same light beam. The light beam 'hits' the active area of a photodiode (PD) (positive intrinsic negative photodiode—PIN PD or an avalanche photodiode—APD) and therefore, the optical signal is converted into an electrical one.

FSO communication, similar to VLC, covers additionally, the ultraviolet (UV) and infrared (IR) spectrum and it doesn't have an illumination requirement, thus, communication links between buildings are a good example of proper application. Using IR for wireless communication over a distance of several kilometres, for a long-range transmission is also referred to as free space optics (FSO). As its first useful applicability, FSO came to bridge the gap between the end user and the already setup fibre optic infrastructure. The ultra-long range, of over 10^4 km solution (also called FSO), has been already successfully applied for Earth-to-satellite or satellite-to-satellite wireless communications [1].

OCC indoor is a practical application of VLC, which uses the camera embedded into smart devices as a transceiver to combine imaging, illumination and wireless communication into a single platform. OCC uses LEDs as transmitters, as laser diodes (LDs) are not appropriate optical transmitters. OCC outdoor is still unreliable since the sunlight direct detection highly interferes with the dedicated optical signal.

Li-Fi, as one of the wireless communication emerging technologies, was coined by Professor Harald Haas in 2011, making an inspired analogy with Wi-Fi and therefore drawing worldwide attention, and thus, raised the interest of both scientific

© Springer Nature Switzerland AG 2021
S. M. Riurean et al., *Application of Visible Light Wireless Communication in Underground Mine*, https://doi.org/10.1007/978-3-030-61408-9_1

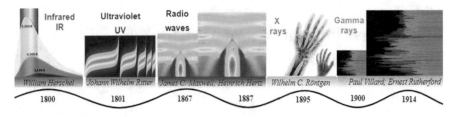

Fig. 1.1 Part of the electromagnetic spectrum discovery timeline

and business communities. Although not standardised or fully developed yet at its entire potential, Li-Fi technology has already been deployed on the market in 2018 and promises to exponentially grow in the near future. Li-Fi can be described as a high speed, fully networked, duplex, multiple input–multiple output (MIMO) wireless communication system that uses the LED lighting fixture and available infrastructure for signal transmission. Li-Fi technology, equivalent to Wi-Fi, transmits data using light instead of radio [2].

Li-Fi concept refers to a system embedded into the LED lighting fixture to allow wireless fast data to be piggybacked by illumination. Full-duplex communication is possible due to download on visible light and upload on the infrared spectrum. Multiple mobile users and wireless handoff from one Li-Fi access point (AP) to another is also possible in early deployed Li-Fi systems.

Since Li-Fi allows multiple gigabits transmission, it '*holds the key to solving challenges faced by spectrum crunch*' and 5G wireless technology due to its strengths: uniquely more secure, virtually interference-free and more reliable than current wireless technologies based on radio frequency (RF) [3].

Furthermore, LEDs lighting is also forecast to replace present illumination light bulbs/lamps (incandescent bulbs, fluorescent or halogen lamps) and spread worldwide due to their obvious advantages (long lifetime—25,000–50,000 h, high energy conversion efficiency, low heat generation, high tolerance to humidity and high/low temperatures, mercury free, compact size). Thus, the market for Li-Fi wireless data communication technology is wide open.

The *IR* systems related to OWC applications use wavelength within the range of 780–950 nm matching the peak sensitivity of low cost, off the shelf photodiodes (PDs). The remote control remains the most widespread application of IR for wireless signal communication.

OWC has the potential to provide security, enough THz of unlicensed bandwidth (BW) and important spatial reuse. The existing light emission diodes (LEDs) lighting infrastructure will soon have simultaneously two functions: illumination and data communication [2].

OWC uses optical wavelengths in the infrared (IR), visible, and ultraviolet (UV) regions of the spectrum. VLC is therefore part of OWC.

Parts of the electromagnetic spectrum (Fig. 1.1) have been discovered and defined chronologically since the eighteenth century, as follows:

– 1800—Sir *William Herschel*—**Infrared IR**

- 1801—*Johann Wilhelm Ritter*—**Ultraviolet UV**
- 1867—*James Clerk Maxwell* anticipated that there should be light with longer wavelengths than IR light
- 1887—*Heinrich Hertz* demonstrated the existence of the waves anticipated by Maxwell by generating *radio waves* in his laboratory
- 1895—*Wilhelm Conrad Röntgen* – **X-rays**
- 1900—*Paul Villard*, investigating radiation from radium, observed **gamma rays**
- 1914—*Ernest Rutherford* coined 'gamma-rays' [4].

Light is an essential support of our everyday life and is becoming an important means of human communication. Most of the authors start their scientific reports regarding the history of wireless optical communication mentioning the ancient times when the light has been used as a reliable means of messages' transmission in the form of smoke signals and beacon fires. For example, long-distance signals were sent by Romans moving polished metallic plates in sunlight. For centuries, after the Great Chinese Wall (21,196 km long) was constructed (200 BC), its guardians were beckoning, announcing Mongolian invasions, with smoke signals.

In the 1790s, semaphore lines were developed based on optical communication systems. Starting from 1792 engineer *Claude Chappe* from France, together with his brother, developed and deployed, until the 1850s, the first optical telegraphy network having 556 stations, covering a total distance of 4800 km, being used for military and national communications [5].

After 1863, the U.S. military used solar power to send information between two mountains' peak (Panamint Ridge and San Gabriel Mountain), creating the wireless solar telegraph, named heliograph, based on Morse code flashes of mirror reflected sunlight [6].

The photophone invented by *Alexander Graham Bell* in June 1880, used sunlight reflected off a selenium photocell and a vibrating mirror to send signals over 200 m [7]. This has been considered as the first application of an FSO communication since Bell succeeded to modulate a voice message onto a light signal [8].

For centuries, blinking lights have been used for navigation, sending messages between ships on sea or communicating with onshore lighthouses.

1.2 Motivation for Using VLC as an Alternative to RF

Data traffic based on wireless communications, mostly indoor, is forecast to exponentially increase in the coming years. This perspective in a short time requires new Gbps type of communication' systems. Today, the main challenge is in developing such wireless systems using high-speed RF, but it is widely acknowledged that it will soon become impossible to be extended because of the limited radio spectrum. Therefore, it is compulsory to explore different parts of the EM spectrum (Fig. 1.2), in order to develop reliable, technologically advanced systems, for wireless communication applications.

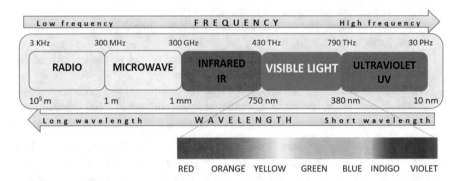

Fig. 1.2 Part of the electromagnetic spectrum with VL emphasised

As history reveals, humankind has been continuously seeking to find a better understanding of life, of physical phenomena in this world, so the light, in all its forms, is not an exception. Researchers have been trying to understand nature in order to use it for a healthier, longer and easier living. One of the most important keys to a better living seems to be human communication. Both face to face and remote, the worldwide human-to-human communication, has always brought significant technological advances.

The last three decades brought many improvements in all aspects of communication, especially in wireless communication, due to the remarkable progress of technology. The Internet is the highest boom ever with the most significant effects in humans' everyday life. The number of the world wide web users is witnessing an incredible growth. According to Internet Worlds Stats, on 31 December 2017, the number of internet users was 4,156,932,140 [9]. It is estimated that, by 2023, there will be 5.3 billion total Internet users (66% of the global population), 3.6 networked devices per capita, 14.7 billion M2M connections, 29.3 billion networked devices [10].

As a result of unprecedented growth of technological advancements in communication, especially in the wireless transmission that allows mobility, about two decades ago the number of IPs (IPv4 has 2^{32}, meaning 4.3 billion addresses) proved to be insufficient for all available devices to be connected to the Internet. The number of logical addresses IPv4 is about to be out of stock, as the Regional Internet Registry (RIR) report (generated at 25 February 2020) shows (Fig. 1.3) [11]. That situation forced the specialists worldwide to find different ways to fulfil the devices' hunger for IPs to be uniquely identified on the Internet. IPv4 has been deployed in 1983 by the Advanced Research Projects Agency Network (ARPANET) as the core protocol of standards based on internetworking methods. More than one decade later, in 1998, the Internet Engineering Task Force (IETF) formalised IPv6 that uses a 128-bit address, allowing in this way a theoretical 3.4×10^{38} available addresses [12]. Therefore, the number of possible addresses IPv6 raises to 2^{128}, more than 7.9×10^{28} times as many as IPv4. There are some transition mechanisms that allow good

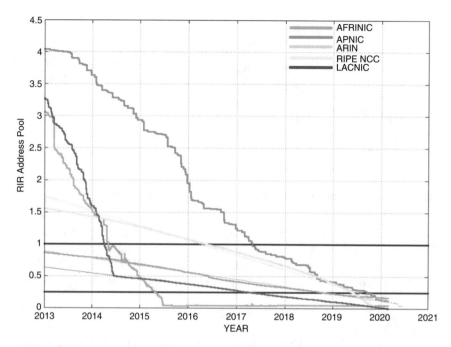

Fig. 1.3 RIR IPv4 Run-Down model [11]. *AFRINIC* African Network Information Centre is the Regional Internet Registry (RIR) for Africa, *APNIC* Regional Internet Registry administering IP addresses for the Asia Pacific, *ARIN* American Registry for Internet Numbers, *RIPE NCC* Regional Internet Registry (RIR) for Europe, the Middle East, and parts of Central Asia, *LACNIC* Regional Internet Registry for the Latin American and Caribbean regions

communication between IPv4 and IPv6 addresses, thus, the unique identification of each smart device connected into the IoT networks seems to be solved, so far.

The RF spectrum extends from 3 kHz to 300 GHz in the range of the EM spectrum and is controlled and used under strict rules. Wi-Fi originally operated in the 2.4 GHz band, then at 5 GHz but now many applications are expected to benefit from the 60 GHz millimetre-wave solutions.

A while ago, the Wireless Gigabit Alliance proposed the use of the millimetre waves in the license-free 60 GHz band, where 7 Gbps short-range wireless links are available for 7 GHz bandwidth. The 60 GHz band is considered as a part of the IEEE 802.11ad structure for very high data throughput in Wireless Local Area Networks (WLANs) using MIMO techniques [13]. Still, the wireless technologies Bluetooth, BLE and Wi-Fi have become victims of their own success. This situation has been created on one hand by a constantly increasing number of devices connected to the Internet and on the other hand by their 'hunger' of data attempting to access higher volumes of multimedia content.

For a short time, millimetre-wave technology can provide a solution to the bandwidth crunch. In this spectrum range, however, links are highly one direction

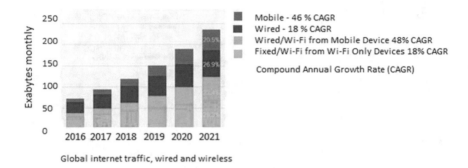

Fig. 1.4 The augmented trend of Wi-Fi domination in Internet traffic [14]

communication, and, therefore, they need complicated digital beamforming and tracking algorithms for their application in mobile wireless networks.

According to the Cisco Visual Networking Index (CVNI) Global IP Traffic Forecast, wired and wireless global internet traffic (Fig. 1.4) doubled in 2019 compared to 2016 [14].

VLC has been identified as a potential solution in order to alleviate the forthcoming RF spectrum crisis.

The visible space of the EM spectrum starts at 380 nm, ends at 750 nm and covers more than 300 THz. It can be used for light communications over a significant wide range of frequencies [15]. VLC can be useful in many communication applications, from millimetres range interconnected within integrated circuits (ICs) up to outdoor kilometres links [16, 17].

Since free Radio Frequency (RF) bandwidth is no longer available worldwide, reliable, partner technologies are to be considered. OWC has lately grown as a highly potential 'reliable partner' of RF communication for an indoor ubiquitous mobile wireless transmission.

However, this spectacular bright side of technological growth brings along a different sight, a difficult to achieve one, that forces scientists to rush the R&D process in order to deploy reliable wireless VLC systems worldwide.

1.3   Significant Worldwide Achievements in Wireless Light-Based Communication

1.3.1   Infrared Wireless Communication

For reliable IR systems, the oTx wavelength must be between 850 and 950 nm and the link lengths up to 10 m. IR wireless communication has today various implementations being categorised depending on the specific application area, the distance between oTx and oRx, and the link type [18].

In the case of *ultra-short-range* IR communication, millimetres links refer to inter and intra chip communications that bring several advantages over copper cabled connections.

Some of the *short-range* IR wireless communication examples are the Wireless Body Area Networks (WBAN) [19] such as wireless medical devices or Wireless Personal Area Networks (WPAN) such as wireless game controllers, computer peripherals, remote controls or remote electronic keys.

Wireless Local Area Networks (WLANs), as *medium range* IR wireless connection, provide indoor IR network connectivity [20].

From 1979 till 2003, significant milestones for IR indoor wireless communication have been established by researchers worldwide (Table 1.1).

1.3.2   Visible Light Communication

2Almost two decades ago (due to important discoveries both regarding the optical and electronic parts of the final devices), the visible field of the electromagnetic spectrum has, for the first time, been considered as a potential candidate for a reliable indoor wireless communication and therefore has been regulated in Japan (Fig. 1.5).

Following the Japanese Visible Light Communication Consortium (VLCC) proposal, in June 2007, the Japan Electronics and Information Technology Industries Association (JEITA) issued the first two visible light standards JEITA CP-1221 and JEITA CP-1222 [25].

The TG7 task group, Institute of Electrical and Electronics Engineers (IEEE), at the beginning of 2009, was working on a VLC standard 802.15.7 covering both Medium Access Control (MAC) and Physical (PHY) layers based on an innovative start. The P802.15.7 IEEE draft standard was published in November 2010 as '*IEEE Standard for Local and Metropolitan Area Networks, Part 15.7: Short-Range Wireless Optical Communication Using Visible Light*' [26].

One of the most important discoveries regarding the type and quality of the main actor of the transmitter device, the LED, has been made by a team of two researchers (*Isamu Akasaki* and *Hiroshi Amano*) from Japanese Nagoya University and an American researcher, *Shuji Nakamura* from University of California. They have been awarded in 2014 the Nobel Prize in Physics for '*the invention of efficient blue light-emitting diodes which has enabled bright and energy-saving white light sources*' [27].

Light fidelity (Li-Fi), the first optical wireless networking technology, was coined by Professor Harald Haas at TED Global talk in 2011 [28]. This talk is considered as one of the milestones in VLC technology since, during his live presentation, professor Haas made both a lexical connection to Wi-Fi and a practical demonstration of the data communication by sending a high definition movie through light using a desk lamp.

Table 1.1 Milestones for IR indoor wireless communication

Year	Researchers	Data rate	Modulation techniques used	Area covered/link range	Original theoretical/technical IR system improvements		
					Optical transmitter (oTx)	Channel	Optical receiver (oRx)
1979	Gfeller and Bapst [21]	64 kbps	Phase-shift keying (PSK)	Range of max 50 m²		• Define physical aspects of the diffuse optical channel	• Baseband PCM receiver • High-pass filter formed by a cascade of 5-RC coupling circuits
		125 kbps	Pulse-code modulation (PCM)				
1996	Kahn and Marsh [22].	50 mbps	On–off keying (OOK)	2.9 m			• Hemispherical concentrator with a hemispherical band-pass optical filter • High-impedance hybrid preamplifier • High-pass filter • Decision-feedback equaliser
1997	Kahn and Barry [23]	50 mbps	OOK, PPM	2–2.5 m	• Quasi-diffuse transmitters	• Baseband channel model • Three non directed NLoS link designs	• Angle diversity receivers
2000	Carruthers and Kahn [24].	70 mbps	OOK, PAM	4.2 m	• Multiple narrow-beam transmitters	• Determine the delay spread in different scenarios	• Multiple, narrow FOV receivers

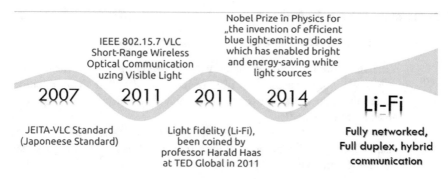

Fig. 1.5 Milestones for VLC standardisation, research and development

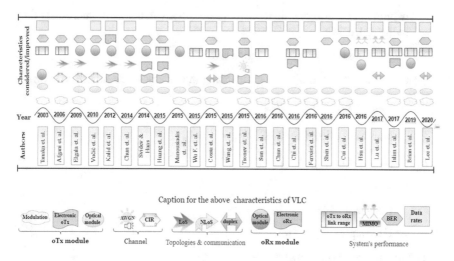

Fig. 1.6 Timeline of data rates and considered/improved characteristics in a VLC system by different researchers worldwide

The main parts of a reliable, consistent VLC indoor system with its key characteristics are (1) the optical transmitter (oTx), (2) the optical channel, (3) topologies and communication mode and (4) the optical receiver (oRx) (Fig. 1.6).

1. The **optical transmitter (oTx)** module refers to:

(a) electrical/electronic part [Electronic oTx],

(b) modulation and data conversion (digital to analogue – DA) and coding [Modulation],

(c) optical module [Optical module]:

– LEDs/picocells/attocells/LDs with their key characteristics.
– particular lenses adapted to a specific environment,

2. The **optical channel** refers to:

    (a) Channel Impulse Response (CIR) ⟨CIR⟩ that depends on:

        – the indoor environmental structure with all the surrounding elements (ceiling, walls, floor), obstacles in front of the oTx with their shape, position, materials and colours necessary to determine their reflective characteristics in order to define CIR,

        – the intrinsic characteristics of the wireless communication optical channel itself (filled with tiny particles of dust, moisture, smoke etc., if any) for the system identification,

    (b) the Additive White Gaussian Noise (AWGN) AWGN consists of different other natural and/or artificial light sources (incandescent, halogen, fluorescent light bulbs or LEDs);

3. Topologies and communication mode refer to:

    (a) the VLC entire setup topology meaning the oTx position related to oRx:

        Line of Sight (LoS) LoS .

        Non Line of Sight (NLoS) NLoS .

    (b) Communication mode can be:

        Simplex.

        Duplex duplex .

        Full duplex.

4. The **optical receiver (oRx)** module refers to:

    (a) optical module Optical module having:

        optical concentrators/lenses,
        filters adapted according to specific optical beam received,
        PIN PD/APD/image sensor with their key characteristics.

    (b) electrical/electronic part Electronic oRx consisting of both hardware with trans impedance amplifier (TIA) and software characteristics related to demodulation, data conversion (analogue to digital) and data decoding.

A VLC's performance is measured by high optical links oTx to oRx link range , bit error ratios ⟨BER⟩, multiple input–multiple output MIMO transmission capability and high data rates Data rates achieved.

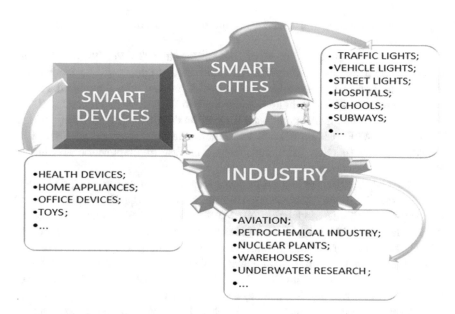

Fig. 1.7 Different domains of application for VLC, IR, OCC and FSO

Different parts of the above-mentioned characteristics have been considered and improved worldwide during the past two decades (Fig. 1.6).

Based on Fig. 1.7, a short literature review regarding the VLC technology, with the main focus on the data rates reached in laboratories worldwide, is presented, underlining the considered VLC characteristics and achieved improvements in each phase.

VLC, as a technology able to convey data indoor using the LED's light, was for the first time demonstrated by *Tanaka* et al. in 2003. The proposed system demonstrated the use of the LED with two simultaneous functions: illumination and data communication. The wireless optical channel model for VLC has been considered, computer simulated and calculated the influence of an optical path difference. Authors had two approaches: first, they used On–Off Keying Return-to-Zero (OOK-RZ) coding to alleviate the optical path delay and second, the Orthogonal Frequency Division Multiplexing (OFDM) coding in order for the delay to be absorbed by a guard interval. For 100 Mbps and 400 Mbps data rates, the results of the Bit Error Ratio (BER) versus received optical power are presented [29].

In 2006, *Afgani* et al. rely on the main advantages of the Intensity Modulation/ Direct Detection (IM/DD), simplicity and low cost, by demonstrating the concept of the transmission scheme Direct Current biased Optical Orthogonal Frequency Division Multiplexing (DCO–OFDM). They took into consideration the main disadvantage of RF communication, high Peak to Average Power Ratio (PAPR) in OFDM and transformed it into an important advantage in VLC by modulating the intensity of a white LED. They proved that the theoretical assumption is very close to experimental results in a practical demonstration of the VLC system in a LoS topology at a distance of 1 m between the optical transmitter (oTx) and optical receiver (oRx) [30].

In 2009 *Elgala* et al. in the paper '*Indoor Broadcasting* via *White LEDs and OFDM*', based on the development of Solid State Lighting (SSL) technology, describe the physical layer application of a VLC system with a modified version of the OFDM modulation technique, and using two Digital Signal Processing (DSP) development boards, made a hardware sample for short-range broadcasting using a White LED (WLED) lamp [31]. It was one of the first references to SSL as its potential duality function for simultaneously the use of illumination and indoor wireless data communication.

By the end of 2010, the highest data rate ever demonstrated in VLC, 513 Mbps was reported by *Vučić* et al. using Discrete Multitone (DMT) modulation and Quadrature Amplitude Modulation (QAM) between a WLED as a source and an APD at the oRx. DMT is based on the idea to split the frequency range (bandwidth) into a large number of frequency bands (channels). Although at a relatively short link length of 300 mm, for an illumination level of about 1000 lx, at a transmission rate of 450 Mbps, the Bit Error Ratio (BER) of the encoded data and a Forward Error Correction (FEC) was demonstrated to be smaller than 2×10^{-3} [32].

The key technologies to achieve high spectral efficiency in optical communication are different FEC techniques and multilevel modulation formats. This combination is known as Coded Modulation (CM), where the sensitivity loss caused by the nonbinary modulation is recovered using FEC [33]. Few months later, in March 2011, the same team from Heinrich-Hertz Institute, Fraunhofer Institute for Telecommunications in Germany, (*Vučić* et al.) almost doubled the speed of VLC data at 803 Mbps using the second important LED, RGB LED, considered in VLC as the main actor of both illumination and data communication. To reach this speed, Wavelength Division Multiplexing (WDM) and DMT modulation were used with DCO-OFDM. Although usually applied in IR, WDM has also been used for the visible light spectrum. WDM is a modulation method that combines multiple signals at various visible light wavelengths for free space transmission. Each wavelength is independently modulated by a set of signals. At the receiver, wavelength sensitive filters as well as visible light colour filters are used [34].

Although in a LoS topology and a 3 cm link, for the first time, the barrier of Mbps has been overpassed at 1 Gbps with a single phosphor-coated WLED, reported by *Khalid* et al. in 2012. The optical source used was a WLED with a 120 Lambertian emission with a luminous flux of about 13 lm at 80 mA bias current. A Lambertian emitter is an optical source having a uniform luminous distribution for all directions [35]. An APD was used with 0.42 A/W responsivity at 620 nm with an active area of 0.8 mm². The discrete multitone waveform, with 512 subcarriers and 180 MHz bandwidth was uploaded into an Arbitrary Waveform Generator (AWG) and then the signal was amplified in order to achieve an appropriate modulation index of the WLED light. Then, via a Bias-Tee, the electrical discrete multitone signal was superimposed on the WLED DC bias current at 80 mA. At the receiver, an integrated Trans Impedance Amplifier (TIA) was used. In front of the APD, was positioned a dichroic optical band pass filter (Semrock type) with 97% transmissivity at 452 and

45 nm Full Width at Half Maximum (FWHM) bandwidth to reject the slow phosphorescent components. A biconvex glass lens has the role of focusing the light onto the APD active area (consisting of 25.4 mm focal length and 25.4 mm diameter) [36].

With a fast polymer colour converter, having a nominal bandwidth (BW) higher than 200 MHz and a Gallium Nitride (GaN) µLED, the data rate of 1.68 Gbps has been achieved in 2014 by a reliable working group of 15 researchers. This time, there are witnessed important efforts on enhancing the oTx/oRx characteristics by optimising the ratio between the blue electroluminescence of the µLED and yellow photo luminescence of the copolymer colour converter [37].

A practical implementation and proof-of-concept of a Spatial Shift Keying (SSK) in a VLC system have been presented by *Videv* and *Haas* in 2014. A matrix of 4 × 4 system LED and PD was used to encode information and decode the spatial signatures (incoming data signal). Applying FEC, the achieved BER was less than 2×10^{-3}. The main challenge in this practical setting of SSK in VLC represents maintaining symbol separation in the received constellation [38].

The highest data rate with the longest transmission distance of 80 cm free space based on an RGB LED, using a pre-equalisation circuit has been achieved by *Huang* et al. in June 2015. They proposed a compact size and easy to install hardware (constant-resistance symmetrical Bridged-T amplitude) equaliser, reaching 1.05 Gbps using Quadrature Amplitude Modulation (64—QAM), OFDM and 1.42 Gbps (7% pre-FEC limit of 3.8×10^{-3}) [39].

A team consisting of 16 British researchers, in 2015, demonstrated a reliable VLC with a data rate of 2.3 Gbps using GaN micro RGB LEDs. With WDM modulation, a higher data rate of 2.3 Gbps in 2015 at a highly dimmed illumination level of just 70 lux was reported [40].

Typically, LEDs can be dimmed either by Constant Current Reduction (CCR) or by Pulse Width Modulation (PWM). In the CCR case, the level of lighting required must be proportional to the current that flows through the LED. Current flows continuously through the LED and is amplified or reduced based on the requirements, suitable for a brighter or dimmed light. PWM dimming refers to the technique that allows switching the current at a high frequency, started from zero to the ranked output current [41].

3.22 Gbps data rate was reported by Chinese researchers *Wu F.M.* et al. based on WDM VLC with a single WLED using Carrier less Amplitude and Phase (CAP) modulation over 25 cm link [42].

Cossu et al. for the first time, in 2015, developed a VLC duplex communication with a length of over 1.5 m into indoor free space with four visible LEDs for 5.6 Gbps downlink and one IR uplink LED for 1.5 Gbps. These promising results for future wide deployment of the system rely on an optimal choice of the optical filter spectra and the LED emission wavelengths. Also, important to underline that all the channels BER were held under the FEC limit of 3.8×10^{-3} [43].

The year 2015 seemed to be a prolific one not only for VLC researchers with many brilliant ideas and implementations but for the VLC technology itself [44], since significant higher data rates have been achieved (8 Gbps) by four Chinese researchers at the end of the year using a hybrid post equaliser and an RGBY LED [45].

According to *Tsonev* et al. as the main disadvantage of WLED consists of the optical efficiency and bandwidth limitation ratio, laser diodes (LDs) are to be considered a potential alternative for better use of VLC for communication purposes. Due to this fact, few off-the-shelf LDs were used by the British researchers in several situations with illumination restraints and the result indicates that optical wireless data rates of 100 Gbps are possible to be reached in usual conditions of standard indoor illumination levels [46].

Chinese researchers from several educational institutions and research facilities fabricated an Aluminium doped Zinc Oxide (AZO) LED with a maximal output power of 42 mW to carry 3 Gbps using 32 QAM and OFDM data at the chip level. They had a different approach to the VLC implementation and proposed to enhance the free space data communication ability of LEDs by improving their non-linearity. A single power type LED, without using any type of pre-equalisation, could reach in their experimental setting, 3 Gbps at high optical power [47].

A new level of data rate was reached by *Chun* et al. in 2016 presented in the paper *'LED Based Wavelength Division Multiplexed 10 Gb/s VLC'*. They used a bit rate adaptive OFDM scheme, to demonstrate a VLC transmission higher than 10 Gbps up to 11.28 Gbps using WDM [48].

During 2016, we noticed that not only high data rates but, very important, a higher distance for communication were achieved. *Chi*, the head of a team consisting of four researchers from China, proposed 8-Pulse Amplitude Modulation (8-PAM) based on Phase-Shifted (PS) Manchester coding using WDM. They reached a data rate of 3.375 Gbps with an RGB LED at a distance higher than 1 m [49].

In June 2016 an important progress in the development of the micro-scale GaN micro-LEDs (µLEDs), optimised for VLC was reported by *Ferreira* et al. They used single pixels from individually addressable arrays of µLEDs having a nominal peak emission wavelength of 450 nm. Wireless data transmission at up to 0.5 m were demonstrated using OOK, 4—PAM and bit loaded OFDM at three different rates: 1.7, 3.4 and 5 Gbps [50].

In September 2016, researchers from Saudi Arabia and the United States published in Optics Express Journal, an article presenting *'the 405 nm emitting Super Luminescent Diode (SLD) with a broad emission of 9 nm at optical power of 20 mW'*. The SLD presented by *Shen* et al. with a 3-dB bandwidth of 807 MHz, reached 1.3 Gbps data rate with OOK modulation [51].

During November 2016, the Journal of Lighting Technology published an investigation of the Multichannel WDM-VLC Communication System. Using a WLED, based on WDM, a VLC channel spacing of 33 nm, the authors *Cui* et al. with 10 channel OOK, achieved up to 5.1 Gbps data streaming [52].

Based on the previous research regarding data rate achieved and distance between LED and PD, at the end of 2016, an imaging MIMO system became available. *Hsu* et al. built a high speed of 3 × 3 imaging MIMO VLC system. The imaging MIMO uses a single lens in front of the PD to focus on three lights. The 3 WLEDs reached

1 Gbps data rate transmission at 1 MHz bit loaded OFDM over a 1-m distance with a high responsive PIN PD and a large spectral response range (340–1040 nm) [53].

Four researchers from Taiwan, at the beginning of 2017, achieved a 6.36 Gbps data rate over 1 m, applying the indoor illumination standard (standard for minimum light level office is 400 lux). Worldwide, illumination standards refer to wellbeing (light characteristics essential for health and comfort), efficiency (light tailored to specific needs), compliance (regulations and specifications comply with environments). *Lu* et al. established a 2 × 2 polarisation-multiplexing MIMO VLC system with RGB LEDs [54].

During March 2017, *Islim* et al. (a team of 15 UK researchers) achieved 11.95 Gbps data rate transmission using 400 nm violet GaN μLEDs, when the source of the VLC's main noise is the nonlinear distortion of the micro-LEDs [55].

A very documented scientific report has been published in Nature Journal in September 2017 that makes a deep investigation of the Violet Laser Diode (VLD) enabled warm white light or daylight with high Colour Rendering Index (CRI) for high-speed lighting communication [56].

A high data rate, the highest ever reached by end of March 2019, for LED based VLC systems of 15.73 Gbps with an FEC coding over a 1.6 m link was achieved by a team of scientists in the University of Edinburgh, using four LEDs (red, blue, green and yellow) and dichroic mirrors [57].

Data rates over 20 Gbps have been demonstrated in 2020 by *Lee C.* et al. in a Li-Fi system using an optic collimator and a laser diode (LD) of surface mount device (SMD) type having 10–100× the brightness of conventional LED sources. The 3 W blue LD used in the experiment offers over 3.5 GHz of 3 dB bandwidth and an SNR above 15 dB up to 1 GHz. QAM and OFDM maximised bandwidth efficiency [58].

As it can be seen in the timeline selection of the most important characteristics regarding the improvements made by numerous authors in a VLC system presented in Fig. 1.6, the highest data rates have been achieved mainly based on the improved key characteristics of optical oTx or oRx, enhanced electronic circuits both of oTx and oRx and more sophisticated modulation techniques.

1.3.3 Optical Camera Communication (OCC)

OCC technology relies on the same emitter as the VLC or IR technology, but instead of a PD (PIN or APD), it uses an optical image sensor. Since optical image sensors are already worldwide embedded into mobile smart devices like smartphones, rear vehicle cameras, digital cameras, surveillance cameras or tablets, one of the most important advantages of OCC is the current off-the-shelf receiver. Most of the image sensors in current cameras are capable to recognise three colours; therefore, RGB LEDs can be used as oTx. OCC has also several drawbacks such as low data rate due to the low sampling rate, out-of-focus effect, unstable frame rate and random block [59].

Although the OCC subject is relatively new, today, more than 150 scientific papers are indexed by the web of science databases. The research community on OCC has proposed different solutions to overcome the above limitations, therefore, due to its multiple advantages, OCC is a valuable candidate for indoor positioning and monitoring, motion capture, IoT and intelligent transportation systems.

Other possible applications of OCC can be dynamic advertising when background LEDs send advertisements to users over a smartphone camera [60]. AR is also a promising application of OCC [61].

1.4   OWC Technologies and Applications

VLC, as well as OCC, have many implementations in different scenarios (Fig. 1.7) such as Indoor Positioning System (IPS), indoor communication (offices, museums, commercial spaces, airplanes, hospitals/healthcare applications, underground transportation, underground mining), mobile connection, vehicle transportation, toys, underwater resource exploration and so on.

Both VLC and OCC systems implemented so far are single way communication in LoS or NLoS topologies or hybrid types. The VL spectrum is used for one way (download for example) and IR or RF is used for the other way (upload) with the same possible topologies (LoS or NLoS). As for the Li-Fi concept, although many improvements and implementations have been noticed, it is not fully developed, yet. Li-Fi consists of a complete wireless network covering bi-directional multiuser transmission, also involving multiple access points. They comprise a wireless network with tiny optical attocells with continuous handover [62].

Due to the last years of intense R&D activity, following the proof of concept, some VLC, IR, OCC, FSO and Li-Fi prototypes have already been implemented in various commercial applications.

The developed technology for wirelessly transmitting and receiving information using light is meant to replace in certain places and situations, the well-known overcrowded and/or forbidden radio communication. Li-Fi uses LEDs to transmit data, which enables data transmission at speed of 10 Gbps in real-world situations. At the same time, some experiments conducted so far suggest that, in certain situations, Li-Fi could be much cheaper to implement than Wi-Fi [63].

1.4.1   Indoor Positioning Systems with VLC and OCC

Most of the Indoor Positioning Systems (IPSs) already well developed are based on different wireless technologies such as Wi-Fi [64], Bluetooth [65], RFID [66], ZigBee [67] or acoustic [68].

On the other hand, many other research papers having the main subject IPS, based on VLC or OCC technologies present a deep survey of the issue with authors all over the world, such as *Do* et al. [69], *Zhang* et al. [70], *Wang* et al. [71], *Yan* et al. [72], *Hassan* et al. [73], *Arafa* [74], *Pisek* [75], *Kim* et al. [76].

A Korean project conducted at a university in Seoul developed an IPS using an Image Sensor (IS) and an Accelerometer Sensor (AS) to provide precise location information. The algorithm proposed in this project (that uses both IS and AS) allows the arbitrary orientation of the mobile device. To improve the precision of the positioning algorithm, a mechanism for image sensor noise decreasing is proposed and then simulated to check the performance of the algorithm [77].

In Europe, Philips Lighting and Osram companies financially supported and deployed IPS in many facilities, mainly retail stores [78].

Due to the many advantages of LEDs, different types of methods—already applied for IPS based on RF are applied in IPS using LEDs—have been proposed, covering different techniques:

- RSSs—Received Signal Strengths [79, 80] with the algorithms: (a) trilateration, (b) fingerprinting and (c) proximity [81];
- AoA—Angle-of-Arrival—has the main disadvantage of high cost but, on the other hand, has a very good accuracy [82] with the algorithms: (a) triangulation, (b) image transformation [81, 83];
- ToA—Time of Arrival—require an accurate synchronisation between the receiver (PD) and emitter (LED) [78] with the algorithms: (a) trilateration and (b) multilateration [81, 84];
- TDoA—Time-Difference-of-Arrival—requires synchronisation between LEDs, and has a high cost for the installation of the positioning system [85] with the algorithms: (a) trilateration, and (b) multilateralism [81];
- Image [85, 86];
- Combination of the above [87].

In 2011, venture-funded start-up company ByteLight [88] launched its indoor location service using VLC and a few years later, in 2015, their patent portfolio together with the intellectual property was acquired by Acuity [89].

The USA company, LVX System [90], during the same period, 2011, claimed to be the first company that launched a VLC commercial product. A few years later, in 2015, LVX System, has entered into a Space Act Agreement (SAA) with NASA to research and develop applications of LVX System's patented VLC technology for use on NASA missions in forthcoming journeys to deep space.

During 2013, a South Korean supermarket uses a guiding light to point out discounts on the smartphone [91]. *Emart*, the app on the smartphone sorts out automatically the lenses and the lights, so shoppers don't need to worry about compatibility issues. Shopping carts have a special place for the smartphone (Fig. 1.8).

This system avoids other technologies like Near Field Communication (NFC) and Quick Response (QR) codes which tend to be either fragmented across devices or too difficult for most of the customers in the supermarket [92].

Lumicast project has been presented in 2016 by the USA's company, *Qualcomm*. They developed a positioning system based on VLC consisting of lighting fixtures of LEDs as the optical transmitter (oTx) and sensors available in commercial smartphone devices as an optical receiver (oRx) (Fig. 1.9).

Fig. 1.8 The Korean Shopping IPS cart with OCC system integrated [86]

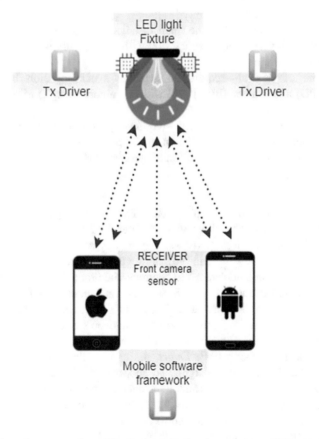

Fig. 1.9 General representation of the Qualcomm's Lumicast project of IPS with OCC (adapted from [87])

Fig. 1.10 The dongle device of the Outstanding Technology company [92]

The system is capable of high position accuracy (length of centimetres) in a tenth of a second. At the same time, it has the ability of device orientation directing and positioning in three dimensions. These capabilities can provide extremely advanced users' experience compared with other available commercial positioning technologies [93].

The smartphone sensor with the support of a mobile software determines the position of the mobile device being able thus to decode the VLC signals. Given its compatibility with existing smartphone devices and LED lighting infrastructure, this technology is able to support a broad range of IPS implementations in offices, commercial places and industrial locations, and has been adopted by leading players in the LED lighting ecosystem such as Acuity Brands [93, 94] and GE Lighting [95].

The Japanese company *Outstanding Technology*, has also launched in 2012 its own product named *Commulight* for wireless communication, achieving high security and Electromagnetic Compatibility (EMC) and high accuracy positioning information platform for smartphones or tablets using LED lighting. Their device (Fig. 1.10) consists of a dongle having a USB or a socket that plugs into any device of 3.5 mm jack and a sensor that senses relevant location-based information from the LED lighting fixtures that transmit data [96].

Commulight is projected to give accurate, real-time information to visitors in a museum, for example, regarding the exhibits, or send coupons to customers in a supermarket or provide precision indoor location services more accurate than Global Positioning System (GPS) or Wi-Fi [97].

Another important developer, *France's Oledcomm* [98], continues to research this technology. Oledcomm claims to be, after years of research, the first company in the world that had deployed Li-Fi products in the market: they equipped in 2012 the first public space, a museum in Europe and deployed in retail (Leclerc) and government-owned facilities like the Paris Metro [99].

In the white paper '*Unlocking the Value of Retail Apps with Lighting*' Philips describes a wall-to-wall IPS based on VLC that enables the United States, the Netherlands and retailers from France to provide personalised location-based services into their store app. Phillips company deployed its first IPS based on VLC in EuraLille Carrefour supermarket on June 2015 [100].

In 2015, at the *Lightfair* in New York, *Acuity Brands* presented a working prototype of indoor location technology based on VLC. VLC proved to be one of the most noticeable topics at fair since a large number of retail stores were interested in applying the new technology in their stores. IPS based on VLC, guide customers to the discounted products they are interested in and send them targeted advertisements and coupons based on the section of the store they are browsing in [101].

1.4.2   Indoor Communication

A subdivision of the Russian company *Stins Coman*, developed in April 2014, RiT Technologies, the wireless optical technology as part of the BeamCaster project launched in order to provide an innovative solution on transferring information to electronic devices and light–medium based on VLC technology [97, 102].

The main part of the network is a light beam attached to a router that is capable to send a signal to a distance of up to 7–8 m. Eight devices transmit the signal at once in other parts of an office on speed four times faster than standard Wi-Fi. According to its developers, the solution has significant advantages due to the mobility and speed of its configuration. Their module transfers data at 1.25 Gbps but they expect enhanced speeds up to 5 Gbps in the near future [94].

According to *Stins Coman* company representatives, their innovation has been tested and implemented in a few countries and different offices for indoor business communication (Fig. 1.11) [102].

An interesting paper published in the US National Library of Medicine, National Institutes of Health magazine, written in 2016 by a team of four researchers from Pakistan, India and the United Kingdom, investigated the previously non-highlighted concept on VLC and LED technologies applied in healthcare [103]. Many other papers investigate the opportunity of embedded VLC applications as well as Li-Fi, both in healthcare facilities and medical devices, as well [104–109].

VLC is a reliable high capacity and radiation free communication system that will be useful in hospitals. Starting from the idea that the lighting fixtures equipped with LEDs are able to use the ubiquitous Power Line Communication (PLC) network that would naturally be able to serve as the backbone network for the Li-Fi and VLC systems in hospitals, *Song* et al. present in their paper an integrated PLC and VLC system with OFDM modulation for the indoor hospital applications, giving a novel solution to replace the conventional wireless communication systems in hospitals [110].

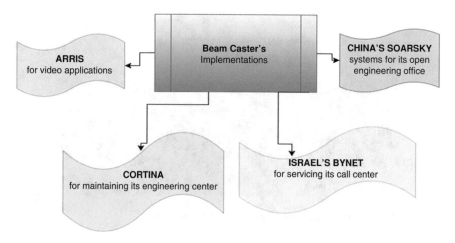

Fig. 1.11 Beam Caster's different implementations (adapted from [102])

Fig. 1.12 Li-Flame Ceiling Unit (CU) (source http://purelifi.com/LiFire/li-flame/)

Fig. 1.13 Li-Flame Desktop Unit (DU) (source http://purelifi.com/LiFire/li-flame/)

1.4.3 Hybrid Indoor Connection

The British company *pureLiFi*, succeeded to launch the first available Li-Fi system, the Li-first, at the 2014 Mobile World Congress in Barcelona [111].

pureLiFi company also launched in 2014 the device Li-Flame consisting of two units: Li-Flame Ceiling Unit (CU) (Fig. 1.12) and Li-Flame Desktop Unit (DU) (Fig. 1.13).

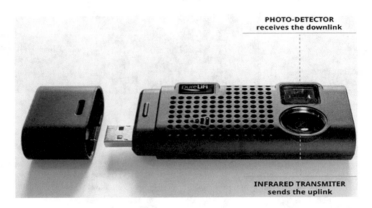

Fig. 1.14 LiFi-X (source: http://purelifi.com/LiFi-products/LiFi-x/)

Li-Flame technology was able to deliver half duplex communication providing 10 Mbps downlink (VLC) and 10 Mbps uplink (IR) up to 3 m with standard LEDs, full mobility (battery-powered, portable desktop unit) with a high data rate. The high data rate can be archived with many Li-Fi Access Points (APs) thus serving multiple users per Li-Fi AP (multiple access) while maintaining high bandwidth for each user.

The state-of-the-art device launched by pureLiFi company, named Li-Fi-X, (Fig. 1.14) has been publicly demonstrated along with the Li-Fi integrated Luminaire at Mobile World Congress in Barcelona on March 2017. The dongle USB-powered station has turned some of the Li-Fi's weaknesses in strengths allowing a full duplex communication with a 40 Mbps downlink with a photodetector as receiver and 40 Mbps uplink with an infrared transmitter. The device allows full mobility being portable and is a multiple users device per Li-Fi Access Point, supported through multiple access.

The company *pureLiFi* also announced signing a contract with Apple, to enable iPhones to use Li-Fi via its on-board camera [3]. *pureLiFi* company underlines that Li-Fi wireless technology, '*holds the key to solving the issues that 5G communication has to encounter*'.

The second generation of *Lucibel* Li-Fi integrated into lighting fixtures has been prepared to be sold on the market [3]. An Academic Evaluation Kit was available for academic research projects at *pureLiFi* company—consisting of Li-Fi-XC station, Li-Fi-XC Access Point and *Lucicup II Luminaires*. The system is possible to be deployed through Power over Ethernet (PoE), a standard main powered solution. In 2018, pureLiFi company has made commercially available a Li-Fi system for the academic community (Fig. 1.15). This system, created in partnership with LED maker *Lucibel*, consists of a modulator that is connected to the lighting fixture (Tx—module) and a USB dongle (Rx module) to connect a display or a computer being Li-Fi-enabled for both illumination and data transfer.

Two young Indian entrepreneurs *Deepak Solanki* and *Saurabh Garg* received funding from BuildIT and therefore launched a start-up venture in Li-Fi technology in Estonia. They developed and launched the '*Jugnu*' project, a new wireless

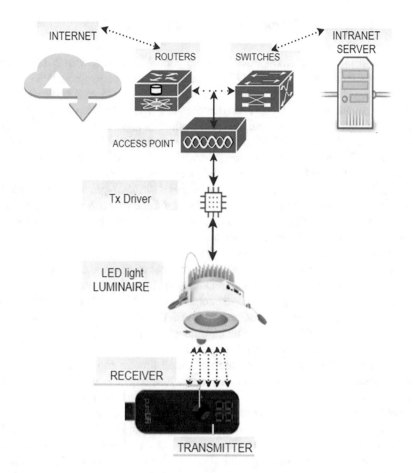

Fig. 1.15 pureLiFi block diagram for the academic evaluation kit: LiF-XC station, LiFi-XC Access Point, Lucicup II Luminaires

technology for high speed up to 1 Gbps data transmission using the VLC concept. *Velmenni* company created VLC links that can transmit data at high speed, send data to up to 20 m, and also works to deploy the system in aviation using its own Li-Fi technology [112].

The project consists of a light wireless entertainment system allowing passengers to access the internet and to wirelessly stream videos and movies through Li-Fi. Airbus considered the weight savings of VLC technology implementation instead of Wi-Fi to be very important since it eliminates wired equipment like seatback screens [113].

A new platform with cybersecurity solutions has been launched on the market by the *Oledcomm* company with the products: MyLiFi®, LiFiMAX (office kit and dongle) and GEO-LiFi® Kit aiming to provide high-speed Internet access with duplex transmission through visible/invisible light [114].

1.4.4   V2V, V2I, I2V and I2I Communication

In the USA Journal, Applied Optics has been published starting with 2012, several interesting papers related to the VLC application in vehicle-to-vehicle (V2V) communication. One of them proposes an early analytical line-of-sight (LoS) path model proved and validated by the measurement results. The influence of both artificial lighting and background solar radiation are studied as important sources of interference when an unobstructed LoS channel has to be guaranteed for this communication system. Few different modulation schemes are analysed in order to evaluate the performance of an outdoor V2V communication based on VL [115].

In 2013 *Căilean* et al. published the paper '*Visible Light Communications: Application to Cooperation between Vehicles and Road Infrastructures*' presenting a first prototype of a VLC system communicating data both between vehicles and between LEDs infrastructure and vehicles.

The red backlight of the vehicle was modulated with OOK amplitude modulation with a microcontroller and a digital power switch that was not too expensive and was used in order to send data to the silicon PD included in the front of a different vehicle behind it up to 15 m away. The experimental results showed that BER was lower than 3×10^{-5} over a distance of maximum of 10 m using a 10 kHz modulation frequency. A four-synchronisation bit configuration has been set with a data length of 4 ASCII characters (4×8 useful bits) [116].

As a result of the unprecedented high accomplishments and based on promising technological capabilities, in Taiwan, the local subsidiary of Ford Motor of the U.S., Ford Lio Ho Motor Co., signed a cooperation agreement with the Department of Computer Science and Information Engineering (CSIE) of the National Taiwan University (NTU) in a vehicular VLC development project. Their main objective was to improve vehicular energy efficiency and lower costs on road infrastructure construction and making driving safer due to an instant react of a vehicle immediately after receiving information from other vehicles [117].

The paper '*A Visible Light Communication based Infra-to-Vehicle Intelligent Transport Demo System*' presented at an international conference by *Fang* et al. addresses a VLC infra-to-vehicle (I2V) system designed and implemented using a specific scenario [118].

The Dubai operator *Du* [119] is one of the companies that demonstrated the Li-Fi's applicable domains, being one of the Middle East's earliest adopters of Li-Fi. *Du*, in cooperation with *Zero.1* (a Dubai-based technology firm), proved how Li-Fi hotspots can be set up to offer internet access [120].

Telecoms operators and technology firms in the United Arab Emirates (UAE) are experimenting with applications for Li-Fi technology [121], hoping that the high-speed data transmission concept will aid the development of smart city systems [122].

1.4.5 VLC Embedded in Toys

Different interesting projects conducted by the Disney Research team in Switzerland presented VLC applications in clothes for children and toys [123] as well as an indoor VLC setup based on networked LEDs [124].

Disney Research has mainly focused on using the visible light to make possible interaction between toys as magic wands or princess dresses [125, 126].

1.4.6 Underwater Resource Exploration Based on VLC Technology

Underwater Optical Wireless Communications (UOWC) research has gained a significant interest during the last few years being a promising alternative method for broadband inexpensive submarine communications, especially due to the fact that has many similarities with the Free Space Optical (FSO) communications or laser links to the satellite. Using the appropriate wavelengths, high data rates can be reached. UOWC of several Mbps has been achieved in laboratory experiments using an aquatic medium that was simulated to have characteristics close to the oceanic waters. It was also demonstrated that UOWC networks are feasible to operate at high data rates for medium distances up to a hundred meters [127].

Kaushal et al. in the paper '*Underwater Optical Wireless Communication*' presents an exhaustive overview of many advances in UOWC by 2016. Channel characterisation, coding techniques and modulation schemes as well as different noise sources specific to UOWC are considered [128].

Using an 80 µm blue emitting GaN based micro-LED, high speed of 800 Mbps data rate was achieved in UOWC by *Tian* et al. at an underwater distance of 0.6 m, with a BER of 1.3×10^{-3}, below the FEC criteria [129].

Wang et al. addressed a hybrid acoustic—optical Underwater Wireless Sensor Network (OA-UWSN) in order to solve the open issue of high-speed communication of real-time images and video in marine information detection. A novel energy-efficient contention based on MAC layer protocol OA-CMAC is also proposed. Based on optical-acoustic fusion technology, in order to achieve high-speed in real-time data communication, the protocol combines the mechanism of Carrier Sense Multiple Access with Collision Avoidance (CSMA/CA) and multiplexing based Spatial Division Multiple Access (SDMA) technology. The proposed MAC protocol was evaluated with OMNeT++ simulations and the results showed that when the optical handshaking success ratio was greater than 50%, it could outcome doubled throughput [130].

1.5 VLC Technology Developed for Underground Mine

Human casualties due to repetitive accidents with unexpected costs in underground mines drove to significant worldwide efforts and numerous research projects dedicated to identifying all the specific risk factors that lead to accidents [131].

Regardless of the underground exploitation, the most frequent accidents hazards are related to inappropriate lighting, rocks or roof falling, structural complexity with narrow spaces and irregular floor or low ventilation level that lead to poor air quality. Poor air quality comes from a high level of dust with different shapes and dimensions of suspended particles in the air, due to exploitation itself or from constant/instant emission of gases specific to mining activity [132]. A low ventilation level results in a denser air with undesired dust particles that would also have a high negative influence on both accuracy and link length of the VLC transmission.

Continuously monitoring and real-time communication of personnel location as well as all type of data regarding risk factors are important issues in underground mining, where the conditions are constantly changing. A fast positioning of personnel (and various equipment operating underground) during normal activity can significantly improve not only security but can also increase daily production, as well.

Different monitoring and communication solutions have already been investigated [133–140] for various mining exploitation. The most studied solutions based on RF communication [141–143] have already been applied in some mining exploitations worldwide.

Therefore, in order to determine the personnel position (and/or various equipment operating underground), the following range-based methods are used for wireless communication technologies using RF, IR or VL spectrum range:

1. Received signal strength (RSSI)—the optical signal of a VLC link slowly becomes weak while the oRx moves away from the light transmitter. In case that the oTx and oRx are tilt and the oTx' optical beam does not meet the oRx, this positioning method is unfeasible.
2. time of arrival (ToA)—uses the time it takes to the EM waves to travel from oTx to oRx. In this method, the oRx's clock is synchronised with the oTx's in order to determine the necessary time for the optical signal to travel from oTx to oRx. However, synchronisation between oTx and oRx is a difficult task,
3. time difference of arrival (TDoA)—for this method, Time Division Multiplexing (TDM) or Frequency Division Multiplexing (FDM) techniques are used. It is assumed that there is a LoS topology between oTx and oRx and at least three sources have to be sensed by the targeted entity (person/equipment) to be localised,
4. angle of arrival (AoA)—used in a LoS topology, both angles (incidence angle and irradiance angle) of the optical signal that hits the active area of the PD are measured. This method needs at least 2 LED lights for 2-D. It is also assumed that the oTx and oRx are aligned in the same X–Y plane, therefore, incidence angles and irradiance angles are equal. Other sensors (accelerometers, for example

[144]) can be used to compensate for the effect of tilt oRxs in combination with multiple PD array. However, using the sensors, the lifetime of the battery in mobile devices is reduced [145].

For example, using Wi-Fi technology, SCADA and Ethernet networks near and far field proximity detection and collision avoidance system as well as underground tracking of personnel and vehicle has been developed by Becker company for Mining in South Africa. One important implementation for miners' security and safety is the underground remote evacuation signalling based on the same technologies mentioned above [146].

However, there are several drawbacks [147] when wireless communication technologies based on RF are used in underground mining:

1. a long response time,
2. the short range of the RF technology (that operates in the 2.4 GHz band) cannot provide large coverage in underground environments [148],
3. RF-based communication is not a suitable solution in some specific underground mine environment where RF signals are limited by strict rules or, their use is even forbidden, as it is the case of the underground coal mining.
4. technologies such as RFID, Wi-Fi or ZigBee involve additional costs for development and particular configuration since they have to be installed in large spaces,
5. low precision (typically between a few tens and several hundreds of meters).

Therefore, alternative solutions have to be searched, developed and implemented [149] to reduce the downsides brought by the use of RF signals but mostly to avoid possible hazards and accidents caused by the signals in the RF spectrum.

VLC is a suitable technology to replace or enhance any underground coal mine communication system, due to some advantages that optical wireless transmission has over the RF communication:

1. LED light is the most suitable type of light to be used in mine for lighting such as in miners' cap lamps but also the luminaires [150].
2. LEDs, as lighting sources used underground are already set into the illumination system on the main galleries underground.
3. the key characteristics of LEDs have been significantly enhanced to be able to convey data at the same time with the illumination,
4. industrial LEDs are also engineered to operate in extremely harsh environments, that is, explosion-proof, resilient to shockwave, immunity to vibration [151–153].

There are different possible architectures and communication modes of the RF, IR and VLC wireless networks [145] dedicated to underground positioning:

1. Remote positioning (RP)—when signals received by oRxs are sent to a central station and the position is determined by combining data received from signal provided by or reflected from the targeted entity (person/equipment) to be localised.

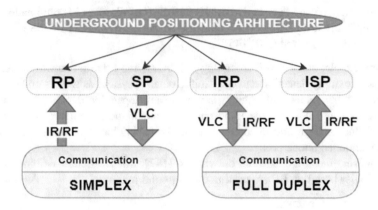

Fig. 1.16 Underground positioning arhitecture and communication modes

2. Self-Positioning (SP)—when the oRx measures the signal from the anchor nodes and then uses the data received to determine its position.
3. Indirect Remote Positioning (IRP)—when an SP oRx sends location data to a remote site or vice versa.
4. Indirect Self-Positioning (ISP)—when the central station sends data about the position to the targeted entity (person/equipment) to be positioned.

The main challenges for reliable underground positioning systems [145] in the case of the architectures mentioned earlier are:

1. RP—when the whole architecture is based on RF technologies, the entire system needs to be enhanced.
2. SP—IR needs fixed PD placed inside the main gallery and data about the targeted entity (person/equipment) to be positioned do not reach the central station.
3. IRP and ISP—a full duplex communication is a challenging task for optical wireless communication (coined in this case, Li-Fi) technology, especially because a MIMO system is expected to be used and therefore intersymbol interference (ISI) has to be mitigated. In this situation, the system complexity and therefore its cost is expected to be high.

No matter the architecture, topology or communication method, the distribution of optical power is sent (by the miner's cap lamp or by luminaire) and received (by luminaire or by the miner's cap lamp) is important to be determined into an underground environment. Depending on the LoS setup, when the miner passes under the luminaire (with LED), the received power could meet the demand of communication [153].

A general representation of the architectures and communication modes of the RF, IR and VLC wireless networks dedicated to underground positioning is presented in Fig. 1.16.

Since the optical signal propagation in underground undergoes many physical phenomena (such as diffraction, reflection, absorption, refraction and scattering) due to the natural composition of such a complex environment (ground, walls and ceiling), very accurate data regarding the optical channel are necessary in order to achieve high-quality optical signal transmission [154].

1.6 Worldwide Cross-Countries Corporation and Funds Invested in VLC Projects

European scientists from Germany, Greece, France, Italy, Austria, Slovenia and the United Kingdom were part of the OMEGA consortium who reached to transfer data at a rate of 100 Mbps using LEDs in the ceiling that light up more than 10 m^2 [155]. Researchers from Heinrich Hertz Institute, Fraunhofer Institute for Telecommunications in Germany, one of the early OMEGA partners, say the receiver can be placed anywhere within this radius, being enough for fast communication. European OMEGA Project was closed in March 2011 and presented several working demonstrations of VLC including high data rates [156].

The UK consortium EPSRC (Engineering and Physical Sciences Research Council) funded by UK universities led by the University of Strathclyde aim to develop Li-Fi innovative technology. Most of the researches around the world concentrate on developing Li-Fi LEDs around 1 mm^2 in size.

The EPSRC team develops tiny, micron-sized LEDs that offer many major advantages: first, the tiny LEDs are able to flicker on and off 1000 times quicker than the larger LEDs, this meaning they would be able to transmit data more quickly. On the other hand, 1000 micron-sized LEDs are fitting into the space of a single larger 1 mm LED. Each of these tiny LEDs action as a separate transmission channel as well as a tiny pixel. Therefore, one large LED array display can also be used as a screen, providing internet connections and displaying information at the same time, as well as room lighting [157].

Students of Mangalore Institute of Technology and Engineering (MITE), in India, demonstrated data communication through the light at high speed and won the third prize at the Annual seventh Edition of UNISYS India Project Competition Cloud 20/20, on April 2016. They proved to be passionate by the technology and drive it to a new level using three colours Triplet Li-Fi (T-Li-Fi), each colour carrying different data streams, thus tripling conventional Li-Fi capacity [158].

In October 2011, the industry group Li-Fi Consortium proved a high-speed optical wireless system and demonstrated the possibility to overcome the limited amount of radio-based wireless spectrum available that is still unexploited. Actually, they are mainly focused on technology development of high data transmission speed. As they demonstrated, this has already been done in relation to different scenarios. They developed new docking technology as well as wireless at high-speed data communication for beaming and provide wireless data broadcasting and a wireless data hotspot [159].

One interesting project, *OpenVLC* aims to develop open-source, software-defined technology, being focused on developing concepts like Li-Fi, built around a credit-card-sized device with a LED front end and embedded Linux platform. It offers a simple physical layer, a set of medium access devices, as well as protocols for the Internet. They also developed a room connector (acting as replicator) that sends data stream between two walls via an optical fibre cable, which connects the two-room connectors on both walls [160].

The Mexico software development company *Sisoft* reached 10 Gbps transfer data ratio across a light spectrum emitted by LED lamps and has caught up with the Scottish research team headed by Harald Haas in achieving 10 Gbps Internet data transmission using VLC also known as Li-Fi, being able to transmit video, audio and data on Internet using light emitted by LED lamps [161, 162]. There are, lately, an increasing number of posts on *GitHub* with Arduino code available for those wishing to build VLC systems as educational projects or at home [163, 164].

Global corporations like *Philips, Toshiba, Samsung, GE, LG, Innotek, Panasonic, Sharp, Cisco, Airbus, Rolls Royce* and *Acuity Brands* (as part of *eldoLED*) are working on Li-Fi, "internet of lights" (IoL) technology and VLC applications for smart cities, as well [160, 161, 165, 166].

As far as we are aware, no financial funds have been invested yet for any kind of underground mining activity based on visible light wireless communication technologies and applications.

References

1. Uysal M., Capsoni C., Ghassemlooy Z., Boucouvalas A. and Udvary, E. (2016). Optical wireless communications—An emerging technology., Signals and Communication Technology, ISBN 978-1-4799-5601-2.
2. Povey, G., *Visible light communications*, http://visiblelightcomm.com [Online]. Last accessed 28.02.2020.
3. [Online] http://purelifi.com/. Last accessed 28.02.2020.
4. Smale, A. (2013). *Discovering the Electromagnetic Spectrum*, High Energy Astrophysics Science Archive Research Center (HEASARC), Astrophysics Science Division (ASD) at NASA/GSFC, https://imagine.gsfc.nasa.gov/science/toolbox/history_ multiwavelength 1. html.
5. Holzmann G. J. and Pehrson B. (1995). The Early History of Data Networks, United States: IEEE Computer Society Press.
6. Woods, D. (2008). Heliograph and mirrors, military communications: From ancient times to the 21st century. In C. Sterling (Ed.), (p. 208). ABC-CLIO. ISBN 978-1851097326.
7. Bell, A. G. (1880). *Bell's Photophone*. Springer Nature. Nov 4, Vol. Nature.

8. Arun, K., & Majumdar, R. J. C. (2007). *Free-space laser communications: Principles and advances*. New York: Springer-Verlag.

9. [Online] https://www.internetworldstats.com/ Last accessed 28.02.2020.

10. [Online] *The Zettabyte era: Trends and analysis—Cisco*. www.cisco.com /c/en/us/ solutions/ collateral/service-provider/visual-networking-index-vni/vni-hyper connectivity-wp. html, Last accessed 28.02.2020.

11. [Online] http://www.potaroo.net/tools/ipv4/index, Last accessed 28.02.2020.

12. Deering, S. and Hinden, R., *Internet protocol, Version 6 (IPv6) specification*. Internet Engineering Task Force (IETF), December 1995. Freely accessible. RFC 1883.

13. [Online] https://www.microwavejournal.com/articles/19079-joint-plugfest-for-wigig-alliance-and-wi-fi-alliance-announced, Last accessed 28.02.2020.

14. [Online] https://newsroom.cisco.com/press-release-content?type=webcontent&article Id=1955935, Last accessed 28.02.2020.

15. Riurean, S., Olar, L. M., Leba, M. and Ionica, A. (2018) *Underground Positioning System Based on Visible Light Communication and Augmented Reality*, Conference Reality Modern Technologies for the 3rd Millennium, Oradea 21–23 March.

16. Dimitrov, S., & Haas, H. (2015). *Principles of LED light communications. Towards networked Li-fi*. Cambridge University Press.

17. Uysal, M., Capsoni, C., Ghassemlooy, Z., Boucouvalas, A., & Udvary, E. (2016). *Optical wireless communications. An emerging technology*. Switzerland: Springer International Publishing. ISBN 978–3–319-30201-0.

18. Carruthers, J. B. (2002). *Wireless infrared communications*. Wiley Encyclopedia of Telecommunications.

19. Riurean, S., Antipova, T., Rocha, Á., Leba, M., Ionica, A., & VLC, O. C. C. (2019). IR and Li-fi reliable optical wireless technologies to be embedded in medical facilities and medical devices. *Journal of Medical Systems, 43*, 308.

20. Khan, J. M., Visible light communication: Applications, architecture, standardization and research challenges, Digital Communications and Networks, Vol. 3, 2, pp. 78–88. ISSN 2352-8648, doi: https://doi.org/10.1016/j.dcan.2016.07.004, (2017).

21. Gfeller, U., & Bapst, R. F. (1979). Wireless in-house data communication via diffuse infrared radiation. *Proceedings of the IEEE, 67*(11), 1474–1486.

22. Kahn, J. M., & Marsh, G. W. (1996). Performance evaluation of experimental 50-Mb/s diffuse infrared wireless link using on–off keying with decision-feedback equalization. *IEEE Transactions on Communications, 44*(11), 1496–1504.

23. Kahn, J. M., & Barry. (1997). Wireless infrared communications. *IEEE Proceedings, 2*, 97.

24. Carruthers, J. B., & Kahn, J. M. (2000). *Angle Diversity for Nondirected Wireless Infrared Communication*. 6, June 2000. IEEE Transactions on Communications, 48, 960–969.

25. Standards, JEITA (2007). *Visible Ligt Communication*. AV&IT Technology Standardisation. https://www.jeita.or.jp/cgi-bin/standard_e/ list.cgi?cateid= 1& subcateid=50 [Online].

26. *P802.15.7/D4—IEEE Draft Standard for Information technology*. s.l.: IEEE, IEEE Xplore Digital Library.35.110—Networking. http://ieeexplore.ieee.org/document/5658207/ Last accessed 28.02.2020, (2010) [Online].

27. https://www.nobelprize.org/nobel_prizes/physics/laureates/2014/press.html Isamu Akasaki, Hiroshi Amano, Shuji Nakamura. The Nobel Prize in Physics 2014. *Nobelprize*. Last accessed 28.02.2020, (2014) [Online].

28. *Wireless data from every light bulb*, TED Talk, Aug 2011, http://bit.ly/tedvlc Last accessed 28.02.2020 [Online].

29. Tanaka, Y., Komine, T., Haruyama, S., & Nakagawa, M. (2003). Indoor visible light data transmission system utilizing White LED lights. *IEICE Transactions on Communications, E86-B*, 2440–2454.
30. Afgani, M., Haas, H., Elgala, H. and Knipp, H. (2006). *Visible Light Communication Using OFDM*, Barcelona, The 2nd International Conference on Testbeds and Research Infrastructures for the Development of Networks and Communities (TRIDENTCOM). pp. 129–134.
31. Elgala, H., Mesleh, R., & Haas, H. (2009). Indoor broadcasting via white LEDs and OFDM. TCE.2009.5277966, s.l. *IEEE Transactions on Consumer Electronics, 55*(3), 1127–1134. IEEE Xplore. https://doi.org/10.1109/TCE.2009.5277966.
32. Vučić, J., Kottke, C., Nerreter, S., Langer, K. D., & Walewski, J. W. (2010). 513 Mbit/s visible light communications link based on DMT-modulation of a White LED, OSA publishing. *Journal of Lightwave Technology, 28*(24), 3512.
33. Alvarado, A., Agrell, E., Lavery, D., and Bayvel, P. (2015). *LDPC Codes for Optical Channels: Is the "FEC Limit" a Good Predictor of Post-FEC BER?* Optical Fiber Communication Conference LoS Angeles, California, United States: OSA Publishing, DOI https://doi.org/10.1364/OFC.2015.Th3E.5.
34. Vučić, J., Kottke, C., Habel, K. and Langer, K.-D. (2011). *Optical wireless network built on white-light LEDs reaches 800Mb/s,* Los Angeles, CA, USA, IEEE Xplore, March, 2011, Optical Fiber Communication Conference and Exposition (OFC/NFOEC) and the National Fiber Optic Engineers Conference.
35. Pedrotti N. and Pedrotti A. (1993). *Introduction to Optics,* Prentice Hall. ISBN 0135015456.
36. Khalid, A. M., Cossu, G., Corsini, R., Choudhury, P., & Ciaramella, E. (2012). 1 Gbit/s visible light communication link based on phosphorescent White LED. *Conference: IEEE Photonics Switching, 4*(2). https://doi.org/10.1109/JPHOT.2012.2210397.
37. Chun, H., Manousiadis, P., Rajbhandari, S., Vithanage, D. A., Faulkner, G., Tsonev, D., McKendry, J. J. D., Videv, S., Xie, E., Gu, E., Dawson, M. D., Haas, H., Turnbull, G. A., Samuel, I. D., & O'Brien, D. C. (2014). *Visible Light Communication Using a Blue GaN μLED and Fluorescent Polymer Color Converter. 20. IEEE Photonics Technology Letters, 26,* 2035–2038. https://doi.org/10.1109/LPT.2014.2345256.
38. Videv, S. and Haas, H. (2014). *Practical space shift keying VLC system*, Conference Location: Istanbul, Turkey, IEEE Xplore Library, Wireless Communications and Networking Conference (WCNC), DOI: https://doi.org/10.1109/WCNC.2014.6952042.
39. Huang, X., Shi, J., Li, J., Wang, Y., & Chi, N. (2015). A Gb/s VLC transmission using hardware Preequalization circuit. *IEEE Photonics Technology Letters, 18*, 1915–1918. https://doi.org/10.1109/LPT.2015.2445781. 15, IEEE Xplore Digital Library, June.
40. Manousiadis, P., Chun, H., Rajbhandari, S., Mulyawan, R., Vithanage, D. A., Faulkner, G., Tsonev, D., McKendry, J. J. D., Ijaz, M., Videv, S., Xie, E., Gu, E., Dawson, M. D., Haas, H., Turnbull, G. A., Samuel, I. D. W., and O'Brien, D. (2015). *Demonstration of 2.3 Gb/s RGB White-light VLC using Polymer based Colour-converters and GaN micro-LEDs,* Nassau, Bahamas, IEEE Summer Topicals Meeting, Visible Light Communications (VisC). https://doi.org/10.1109/PHOSST.2015.7248279.
41. *How do LEDs work.* Philips Lighting. Philips. http://www.lighting.philips.com/main/education/lighting-university/lighting-university-browser/video/LEDs. Last accessed 28.02.2020 [Online].
42. Wu, F., Lin, C., Wei, C., Chen, C., Chen, Z. and Huang, K. (2013). *3.22-Gb/s WDM visible light communication of a single RGB LED employing carrier-less amplitude and phase modulation,* in Optical Fiber Communication Conference/National Fiber Optic Engineers Conference, OSA Technical Digest (online) (Optical Society of America, 2013), paper OTh1G.4 pp. 1–3.
43. Cossu, G., Ali, W., Corsini, R., & Ciaramella, E. (2015). Gigabit-class optical wireless communication system at indoor distances (1.5–4 m). *Optical Society of America.*, OSA Publishing, *23*(12), 15700. https://doi.org/10.1364/OE.23.015700.

44. Chi, Y.-C., Huang, Y.-F., Wu, T.-C., Tsai, C.-T., Chen, L.-Y., Kuo, H.-C., & Lin, G.-R. (2017). Violet laser diode enables lighting communication, s.l. *A Natural Research Journal.* https://doi.org/10.1038/s41598-017-11186-0.
45. Wang, Y., Tao, L., Huang, X., Shi, J., & Chi, N. (2015). 8-Gb/s RGBY LED-based WDM VLC system employing high-order CAP modulation and hybrid post equalizer. *Photonics Journal, IEEE, 7*(6), 1–7.
46. Tsonev, D., Videv, S., & Haas, H. (2014). Towards a 100 Gb/s visible light wireless access network, optics express, OSA publishing, 2015. *Optics Express, 23*(2), 1627–1637. https://doi.org/10.1364/OE.23.001627.
47. Shen, C., Lee, C., Ng, T. K., Nakamura, S., & Speck, J. S. (2016). High-speed 405-nm superluminescent diode (SLD) with 807-MHz modulation bandwidth, OSA publishing. *Optics Express, 24*(18), 20281–20286.
48. Cui, L., Tang, Y., Jia, H., Luo, J., & Gnade, B. (2016). *Analysis of the Multichannel WDM-VLC Communication System,* OSA publishing. *Journal of Lightwave Technology, 34* (24), 5627–5634.
49. Chin-Wei, H., Chi-Wai, C., Cheng, L. I., Yen-Liang, L., Yeh, C.-H., & Yang, L. (2016). High speed imaging 3 × 3 MIMO phosphor White-light LED based visible light communication system. *IEEE Photonics Journal, 8*(6). https://doi.org/10.1109/JPHOT.2016.2633395.
50. Lu, I. C., Lai, C. H., Yeh, C. H., & Chen, J. (2017). *6.36 Gbit/s RGB LED-based WDM MIMO visible light communication system employing OFDM modulation.* Los Angeles, CA.: OSA Technical Digest, Optical Society from America, March, Optical Fiber Communication Conference. https://doi.org/10.1364/OFC.2017.W2A.39.
51. Islim, M. S., Ferreira, R. X., He, X., Xie, E., Videv, S., Viola, S., Watson, S., Bamiedakis, N., Penty, R. V., White, I. H., Kelly, A. E., Gu, E., Haas, H., & Dawson, M. D. (2017). Towards 10 Gb/s orthogonal frequency division multiplexing-based visible light communication using a GaN violet micro-LED, OSA publishing. *Photon Research, 5*(2), A35–A43. https://doi.org/10.1364/PRJ.5.000A35.
52. Elgala, H., et al. (2016). Coexistence of WiFi and Li-fi toward 5G: Concepts, opportunities, and challenges. *IEEE Communications Magazine, 54*(2), 64–71.
53. Bian, R., Tavakkolnia, I., & Haas, H. (2019). 15.73 Gb/s visible light communication with off-the-shelf LEDs. *Journal of Lightwave Technology, 37*(10), 2418–2424. https://doi.org/10.1109/JLT.2019.2906464.
54. Lee, C., Islim, M. S., Videv, S., Sparks, A., Shah, B., Rudy, P., McLaurin, M., Haas, H., & Raring, J. (2020). Advanced LiFi technology: Laser light. *Proc. SPIE, Light-Emitting Devices, Materials, and Applications, XXIV,* 1130213. https://doi.org/10.1117/12.2537420.
55. Ayyash, M., Elgala, H., et al. (2016). Coexistence of WiFi and Li-fi toward 5G: Concepts, opportunities, and challenges. *IEEE Communications Magazine, 54*(2), 64–71. https://doi.org/10.1109/MCOM.2016.7402263.
56. Arnon, S. (2015). *Visible light communication.* Cambridge University Press.
57. Ghassemlooy, Z., Luo, P., & Zvanovec, S. (2016). In M. Uysal et al. (Eds.), Optical Wireless Communications, Signals and Communication Technology *Optical camera communications. Signals and communication technology* (pp. 547–568). Switzerland: Springer International Publishing. https://doi.org/10.1007/978-3-319-30201-0_25.
58. Tsonev, D., Videv, S., & Haas, H. (2013). Light Fidelity (Li-fi): Towards all-optical networking. *Proceedings of SPIE—The International Society for Optical Engineering, 9007,* 900702. https://doi.org/10.1117/12.2044649.
59. He, S., & Chan, S. G. (2016). Wi-fi fingerprint-based indoor positioning: Recent advances and comparisons. *IEEE Communications Surveys & Tutorials, 18*(1), 466–490. Firstquarter.
60. Hossain, A. K. M. M., Soh, W. S. (2007). *A comprehensive study of bluetooth signal parameters for localization,* in: IEEE 18th Int. Symp. on Personal, Indoor and Mobile Radio Commun., pp. 1–5.

61. Yang, P., & Wu, W. (2014). Efficient particle filter localization algorithm in dense passive RFID tag environment. *IEEE Transactions on Industrial Electronics, 61*(10), 5641–5651.
62. Konings, D., Faulkner, N., Alam, F., Noble, F., Lai, E. M.-K. (2017), *The Effects of Interference On The RSSI Values of a ZigBee based Indoor Localization System*, 24th International Conference on Mechatronics and Machine Vision in Practice (M2VIP), Auckland, New Zealand.
63. Erol-Kantarci, M., Mouftah, H. T., & Oktug, S. (2011). A survey of architectures and localization techniques for underwater acoustic sensor networks. *IEEE Communications Surveys and Tutorials, 13*(3), 487–502.
64. Do, T.-H., & Yoo, M. (2016). An in-depth survey of visible light communication based positioning system. *Sensors, 16*, 678.
65. Zhang, W., & Kavehrad, M. (2013). *Comparison of VLC-based indoor positioning techniques.* San Francisco, CA: SPIE OPTO, International Society for Optics and Photonics. https://doi.org/10.1117/12.2001569.
66. Wang, C., et al. (2012). *The research of indoor positioning based on visible light communication* (Vol. 12, pp. 85–92). China: Communication., 2015.
67. Yan, K., et al. (2015). *Current status of indoor positioning system based on visible light,* In Proceedings of the 15th IEEE International Conference on Control, Automation and Systems (ICCAS), Busan, Korea, October, pp. 565–569.
68. Hassan, N. U., Naeem, A., Pasha, M. A., Jadoon, T., & Yuen, C. (2015). Indoor positioning using visible LED lights: A survey, s.l. *ACM Computing Surveys, 48*, 20.
69. Arafa, A. T. (2015). *An Indoor Optical Wireless Location Comparison between an Angular Receiver and an Image Receiver,* Doctoral Dissertation, University of British Columbia, Vancouver, Canada.
70. Pisek, E., Rajagopal, S. and Abu-Surra, S. (2012). *Gigabit Rate Mobile Connectivity Through Visible Light Communication*, IEEE International Conference on Communications (ICC) Ottawa, ON, Canada, doi: https://doi.org/10.1109/ICC.2012.6363739.
71. Kim, H.-S., Kim, D.-R., Yang, S.-H., Son, Y.-H., & Han, S.-K. (2013). An indoor visible light communication positioning system using a RF carrier allocation technique. *Journal of Lightwave Technology*, 134–144. Standard No. 31.
72. Huynh, P., & Yoo, M. (2016). VLC-based positioning system for an indoor environment using an image sensor and an accelerometer sensor, s.l. *Image Sensor Based Optical Wireless Communications, 16*(6), 783.
73. Halper, M. (2017). Contributing Editor, LEDs Magazine, and Business/Energy/Technology Journalist. *Two more indoor positioning projects sprout in European supermarkets.* https://www.ledsmagazine.com 8 March [Online].
74. Mousa, F. I. K., et al. (2016). Indoor localization system utilizing two visible light emitting diodes. *Optical Engineering, 55*(11), 114–116.
75. Kim, H. S., Kim, D. R., Yang, S. H., Son, Y. H., & Han, S. K. (2012). An indoor visible light communication positioning system using a RF carrier allocation technique. *Journal of Lightwave Technology, 31*(1), 134–144.
76. Zhuang, Y., et al. (2018). A survey of positioning systems using visible LED lights. *IEEE Communications Surveys & Tutorials, Third Quarter, 20*(3), 1963–1988. https://doi.org/10.1109/COMST.2018.2806558.
77. Eroglu, Y. S., Guvenc, I., Pala, N. and Yuksel, M. (2015). *AOA-based localization and tracking in multi-element VLC systems,* Proceedings Wireless Microwave Technology Conference, pp. 1–5.
78. Wang, T. Q., Sekercioglu, Y. A., Neild, A., & Armstrong, J. (2013). Position accuracy of time-of-arrival based ranging using visible light with application in indoor localization systems. *Journal of Light Technology, 31*, 3302–3308.
79. Nadeem, U., Hassan, N. U., Pasha, M. A., & Yuen, C. (2014). Highly accurate 3D wireless indoor positioning system using white LED lights. *Electronic Letters, 50*(11), 828–830.

80. Kuo, Y. S., Pannuto, P., Hsiao, K. J. and Dutta, P. (2014). *Luxapose: Indoor positioning with mobile phones and visible light,* Proc. 20th Annu. Int. Conf. Mobile Comput. Netw, pp. 447–458.

81. Rahman, M. S., & Kim, K. D. (2013). Indoor location estimation using visible light communication and image sensors. *International Journal of Smart Home, 7,* 166–170.

82. Yang, S. H., Kim, H. S., Son, Y. H., & Han, S. K. (2014). Three-dimensional visible light indoor localization using AOA and RSS with multiple optical receivers. *Journal Lightwave Technology, 32*(14), 2480–2485.

83. Wright, M. (2015). *Acuity acquires indoor-location-services specialist ByteLight.* http://www.ledsmagazine.com/articles/2015/04/acuity-acquires-indoor-location-service-specialist-bytelight.html. [Online].

84. http://www.ledsmagazine.com/ articles/2015/04/acuity-acquires-indoor-location-service-specialist-bytelight.html]. (2015) [Online].

85. http://www.lvxsystem.com/. [Online] Last accessed 28.02.2020.

86. Millward, S. (2013). https://www.techinasia.com/korean-supermarket-emart-led-lights-smartphone-app-discounts. 17 April [Online]. Last accessed 28.02.2020.

87. Jovicic, A. *A high accuracy indoor positioning system based on visible light communication,* Whitepaper https://pdfs.semanticscholar.org/69da/67e63fa2ae0b771819916adf41817e40cd59.pdf. Last accessed 28.02.2020 [Online].

88. Press-release, *Qualcomm and Acuity Brands collaborate to commercially deploy Qualcomm Lumicast Technology for precise indoor location services in more than 100 retail locations,* https://www.qualcomm.com/news/releases/2016/03/14/qualcomm-and-acuity-brand. Last accessed 28.02.2020 [Online].

89. Lydecker, S. *Illuminating the in-store experience.* Indoor positioning white paper revised110315.pdf. http://www.acuitybrands.com/solutions/–/media/files/acuity/solutions/services/bytelight services indoor positioning. Last accessed 28.02.2020 [Online].

90. *GE intelligent lighting to transform retail experience through Qualcomm Collaboration,* http://pressroom.gelighting.com/news/ge-intelligent-lighting-to-transform-retail-experience-through-qualcomm-collaboration. May 2015. Last accessed 28.02.2020 [Online].

91. http://www.ot-c.co.jp/ Last accessed 28.02.2020 [Online].

92. Gorman, M.. https://www.engadget.com/2012/07/16/outstanding-technology-visible-light-communication, Last accessed 28.02.2020 [Online].

93. http://www.oledcomm.com, Last accessed 28.02.2020 [Online].

94. http://www.slate.fr/story/104255/Li-Fi-transmission-donnees-lumiere. Last accessed 28.02.2020 [Online].

95. Kelion. BBC News. Technology. *Supermarket LED lights talk to smartphone app.,* 22 May 2015. http://www.bbc.com/news/technology-32848763. Last accessed 28.02.2020 [Online].

96. McGrath, D. (2015). *Retailers test visible light communications. Electronics 360* http://electronics360.globalspec.com/article/5360/retailers-test-visible-light-communications. Last accessed 28.02.2020 [Online].

97. http://rbth.com/science_and_tech.russian_firms_Li-Fi_internet_solution_winning_foreign_client_37805.html, https://www.rbth.com/science-and-tech. 30 June 2016, Last accessed 28.02.2020 [Online].

98. Muhammad, S., Qasid, S. H. A., Rehman, S., & Rai, A. B. S. (2016). Visible light communication applications in healthcare. *Technology and Health Care, 24*(1), 135–138. https://doi.org/10.3233/THC-151098.

99. Cahyadi W. A., Jeong T. I, Kim Y. H., Chung Y. H. and Adiono T. (2015). *Patient monitoring using visible light uplink data transmission,* Proceedings of International Symposium on Intelligent Signal Processing and Communication Systems, ISPACS, pp. 431–434.

100. Ding, W., Yang, F., Yang, H., Wang, J., Wang, X., et al. (2015). A hybrid power line and visible light communication system for indoor hospital applications. *Computers in Industry, 68,* 170–178.

101. An J. Y. and Chung W. Y. (2016). *Bio-medical data transmission system using multi-level visible light based on resistor ladder circuit*, JSST25, pp. 131–137.
102. Riurean, S. M., Leba, M., & Ionica, A. (2019). VLC embedded medical system architecture based on medical devices quality requirements. Iss. S1, Bucharest. *Journal Quality-Access to Success, 20*(1), 317.
103. Riurean, S., Antipova, T., Rocha, A., Leba, M, Ionica, A. (2019). *Li-Fi Embedded Wireless Integrated Medical Assistance System,* 16–19 April 2019. WorldCist'19—7th World Conference on Information Systems and Technologies Spain.
104. Ali, H., Ahmad, M. I. and Malik, A. (2019). *Li-Fi Based Health Monitoring System for Infants,* 2nd International Conference on Communication, Computing and Digital systems (C-CODE), Islamabad, Pakistan, pp. 69–72. doi: https://doi.org/10.1109/C-CODE.2019.8681012.
105. Song, J., et al. (2014) *Indoor hospital communication systems: an integrated solution based on power line and visible light communication.* Monaco, 4–6 May 2014, Faible Tension Faible Consommation.
106. *pureLiFi to demonstrate first ever Li-Fi system* at Mobile World Congress Virtual-Strategy Magazine. 19 February 2014. Last accessed 28.02.2020 [Online].
107. [https://yourstory.com/2015/05/velmenni/] [http://www.fiercetelecom.com/ telecom /estonias-velmenni-to-release-Li-Fi-broadband-led-bulbs-2018-19.   Last   accessed   28.02.2020 [Online].
108. Flynn, D. (2016). https://www.ausbt.com.au/airbus-wants-to-upgrade-wifi-to-the-speed-of-light. Last accessed 28.02.2020 [Online].
109. https://www.oledcomm.net/lifimax1g-industrials-iot/ Last accessed 28.02.2020 [Online].
110. Cui, K., Chen, G., Xu, Z., & Roberts, R. D. (2012). *Traffic light to vehicle visible light communication channel characterization.* S.l. *OSA Publishing Applied Optics, 51,* 6594–6605. https://doi.org/10.1364/AO.51.006594.
111. Căilean, A., Cagneau, B., Chassagne, L., Topsu, S., Alayli, Y., Blosseville, J. M. and de Henares, A. (2012). *Visible light communications: application to cooperation between.* Spain: Intelligent Vehicles Symposium (IV2012).
112. Liang, Q. and Ford, N. T. U. *Tie Up in VLC Project to Enhance Driving Safety.* 2015. http://www.cens.com/cens/html/en/news/news_inner_48751.html,   Last   accessed   28.02.2020 [Online].
113. Fang, P., Bao, Y., Shen, J. and Chen, J. (2015). *A Visible Light Communication based Infra-to-Vehicle Intelligent Transport Demo System.* Shenzhen, China: IEEE Xplore, International Conference on Connected Vehicles and Expo. doi: https://doi.org/10.1109/ICCVE.2015.62.
114. http://www.du.ae/. Last accessed 28.02.2020 [Online].
115. http://www.zero1.zone/. Last accessed 28.02.2020 [Online].
116. http://whatis.techtarget.com/definition/Li-Fi. Last accessed 28.02.2020 [Online].
117. http://www.thenational.ae/business/telecoms/end-of-wi-fi-in-the-uae-du-trials-super-speedy-Li-Fi. Last accessed 28.02.2020 [Online].
118. Schmid, S., Gorlatova, M., Giustiniano, D., Vukadinovic, V. and Mangold, S. (2015). *Networking Smart Toys with Wireless ToyBridge and ToyTalk,.* s.l.: Poster Session Infocom 2011 Springer International Publishing.
119. Schmid, S., Richner, T., Mangold, S., Thomas, B., and Gross, R. (2016). *EnLighting: an indoor visible light communication system based on networked light bulbs.* Sensing, Communication, and Networking (SECON), 13th Annual IEEE International, Switzerland.
120. Schmid, S., Corbellini, G., Mangold, S. Gross, T. and Anaheim R. (2012). *An LED-to-LED visible light communication system with software-based synchronization.* California, USA, Dec. 3–7. Globecom Workshops. doi: https://doi.org/10.1109/GLOCOMW.2012.6477763.
121. Corbellini, G., Aksit, K., Schmid, S., Mangold, S., & Gross, T. R. (2014). Connecting networks of toys and smartphones with visible light communication. *IEEE Communications Magazine, 52,* 72–78.
122. Gkoura, L., Roumelas, G., Nistazakis, H. E. and Tombras, G. S. (2017). *Underwater optical wireless communication systems: A concise review.* July. doi: https://doi.org/10.5772/67915.

123. Kaushal, H., & Kaddoum, G. (2016). *Underwater optical wireless communication.* s.l. *IEEE Access, 4,* 1518–1547. https://doi.org/10.1109/ACCESS.2016.2552538.
124. Tian, P., Liu, X., Yi, S., Huang, Y., Zhang, S., Zhou, X., Hu, L., Lirong, Z., & Liu, R. (2017). High-speed underwater optical wireless communication using a blue GaN-based micro-LED. *Optics Express, 25,* 2. https://doi.org/10.1364/OE.25.001193.
125. Wang, J., Shen, J., Shi, W., Qiao, G., Wu, S., & Wang, X. (2019). A novel energy-efficient contention-based MAC protocol used for OA-UWSN. s.l. *Sensors (Basel), 19,* 1. https://doi.org/10.3390/s19010183.
126. Ullah, M. F., Alamri, A. M., Mehmood, K., Akram, M. S., Rehman, F., Rehman, S. U., & Riaz, O. (2018). Coal mining trends, approaches, and safety hazards: A brief review. *Arabian Journal of Geosciences, 11,* 1–16.
127. Khanzode, V. V., Maiti, J., & Ray, P. K. (2011). A methodology for evaluation and monitoring of recurring hazards in underground coal mining. *Safety Science, 49*(8–9), 1172–1179.
128. Forooshani, A. E., Bashir, S., Michelson, D. G., & Noghanian, S. (2013). A survey of wireless communications and propagation Modeling in underground mines. *IEEE Communications Surveys and Tutorials, 15,* 1524–1545.
129. Jiping, S., & Chenxin, L. (2014). Mine TOA location method based on Kalman filter and fingerprinting. *Journal of China University of Mining & Technology, 43*(6), 1127–1133.
130. Wang, L. (2015). *Research on key algorithm for underground personnel location based on pedestrian dead reckoning.* China University of Mining and Technology.
131. Lu, Q., Liao, X., Xu, S., et al. (2016). *A hybrid indoor positioning algorithm based on WiFi fingerprinting and pedestrian dead reckoning,* IEEE.
132. Shiyin, L., Wang, H., & Nan, Z. (2017). Underground personnel location system based on MEMS inertial sensor. *Coal Mine Safety, 48*(4), 111–114.
133. ZheXing, S., & Wang, Y. W. (2018). Accurate two-dimensional location method for mine personnel based on Kalman filter. *Mining and Industry Automation, 44*(05), 31–35.
134. Wang, J., Guo, Y., Guo, L., et al. (2019). Performance test of MPMD matching algorithm for geomagnetic and RFID combined underground positioning. *IEEE Access, 7,* 129789–129801. https://doi.org/10.1109/ACCESS.2019.2926098.
135. Riurean, S., Olar, M., Leba, M., Ionica, A. (2018). *Underground positioning system based on visible light communication and augmented reality,* 17th International Technical-Scientific Conference on Modern Technologies for the 3rd Millennium, Oradea, Romania, Modern Technologies for the 3rd Millennium, p. 345–350.
136. Hadenius, P. (2006). Underground Wi-fi—cities may wait, but mines get full wireless broadband coverage. *Technology Review, 109,* 1.
137. Sun, H., Bi, L., Lu, X., et al. (2017). Wi-fi network-based fingerprinting algorithm for localization in coal mine tunnel. *Journal of Internet Technology, 18*(4), 731–741.
138. Mohapatra, A. G., et al. (2018). Precision local positioning mechanism in underground mining using IoT-enabled WiFi platform. *International Journal of Computers and Applications.* https://doi.org/10.1080/1206212X.2018.1551178.
139. Yasir, M., Ho, S.-W., & Vellambi, B. N. (2016). Indoor position tracking using multiple optical receivers. *Journal of Lightwave Technology, 34*(4), 1166–1176.
140. Seguel, F., Soto, I., Adasme, P., & Charpentier, P. (2017). *Potential and challenges of VLC based IPS in underground mines, first south American colloquium on visible light communications.* Chille: Santiago. https://doi.org/10.1109/SACVLC.2017.8267610.
141. https://www.becker-mining.com/en/news/south-africa/becker-mining-wi-fi-system-which-ensures-reliable-safe-and-efficient-communication, Last accessed 28.02.2020 [Online].
142. Firoozabadi, A. D., et al. (2019). A novel frequency domain visible light communication (VLC) three-dimensional trilateration system for localization in underground mining. *Applied Sciences, 9,* 1488. https://doi.org/10.3390/app9071488.
143. Ferrer-Coll, J., Angskog, P., Shabai, P., Chilo, J., Stenumgaard, J. *Analysis of wireless communications in underground tunnels for industrial use.* In Proceedings of the 38th Annual

Conference on IEEE Industrial Electronics Society (IECON), Montreal, QC, Canada, 25–28 October 2012.

144. Iturralde, D., et al. (2014). *A new location system for an underground mining environment using visible light communications*, Networks & Digital Signal Processing (CSNDSP), 9th International Symposium on. IEEE.

145. Wang, Y., Chi, N., Wang, Y., et al. (2015). Network architecture of a high-speed visible light communication local area network. *Journal of Photonics Technology Letters IEEE, 27*(2), 197–200.

146. Yenchek, M. R., & Sammarco, J. J. (2010). The potential impact of light emitting diode lighting on reducing mining injuries during operation and maintenance of lighting systems. *Safety Science, 48*(10), 1380–1386.

147. Krommenacker N., Vasquez O. C., Alfaro M. D., Soto I. (2016). *A self-adaptive cell-ID positioning system based on visible light communications in underground mines*, IEEE International Conference on Automatica (ICA-ACCA), doi: https://doi.org/10.1109/ICA-ACCA.2016.7778427.

148. Wua, G., Zhangb, J. (2016). *Demonstration of a visible light communication system for underground mining applications*, International Conference on Information Engineering and Communications Technology (IECT 2016) ISBN: 978–1–60595-375-5.

149. Zaarour, N., Kandil, N., Hakem, N. and Despins, C. (2012). *Comparative experimental study on modeling the path loss of an UWB channel in a mine environment using MLP and RBF neural networks*, International Conference on Wireless Communications in Underground and Confined Areas, Clermont Ferrand, pp. 1–6. doi: https://doi.org/10.1109/ICWCUCA.2012.6402503149.

150. OMEGA project: Home Gigabit Access project, www.ictomega.eu, (2008). Last accessed 28.02.2020 [Online].

151. Grallert, H. J.and Boche, H. (2007). *Innovations for the digital future* Annual Report [Online].

152. https://www.epsrc.ac.uk, Last accessed 28.02.2020 [Online].

153. http://www.thehindu.com/news/national/karnataka/city-students-win-praises-for-triple-Li-Fi-project/article8444278.ece. Last accessed 28.02.2020 [Online].

154. http://lificonsortium.org/speed.html. Last accessed 28.02.2020 [Online].

155. Wang, Q., Giustiniano, D. and Puccinelli, D. (2014). *OpenVLC: Software-defined visible light embedded networks,* 978–1–4503-3067-1. pp. 15–20.

156. Vega, A. *Li-Fi record data transmission of 10 Gbps set using LED lights.* Engineering and Technology Magazine. Retrieved 29 November 2015. Last accessed 28.02.2020 [Online].

157. http://www.ledinside.com/news/2014/7/mexican_software_company_sisoft_li_fi_transmis sion_reaches_10gbps. (2014). Last accessed 28.02.2020 [Online].

158. *GitHub.* https://github.com/jpiat/arduino/wiki/Arduino-simple-Visible-Light-Communication Last accessed 28.02.2020 [Online].

159. *GitHub.* https://github.com/c-i-a-n-i/Final-year-Li-Fi-Project. Last accessed 28.02.2020 [Online].

160. Leba, M., Riurean, S. and Ionica, A. (2017). *Li-Fi—The path to a new way of communication,* Lisbon, 12th Iberian Conference on Information Systems and Technologies (CISTI), pp. 1–6. doi: https://doi.org/10.23919/CISTI.2017.7975997.

161. Vappangi, S., & Vakamulla, V. M. (2018). Synchronization in visible light communication for smart cities. *IEEE Sensors Journal, 18*(5), 1877–1886. https://doi.org/10.1109/JSEN.2017.2777998.

162. www.internetworldstats.com/stats.htm. Last accessed 28.02.2020 [Online].

163. www.apnic.net/community/ipv4-exhaustion/ipv4-exhaustion-details/. Last accessed 28.02.2020 [Online].

164. Cossu, Y., Ali, W., Corsini, R., & Ciaramella, E. (2015). Gigabit-class optical wireless communication system at indoor distances (1.5–4 m). *Optics Express, 23*, 15700–15705.

165. Sun, Z., Teng, D., Liu, L., Huang, X., Zhang, X., Sun, K., Wang, Y., & Chi, N. (2016). A power-type single GaN-based blue LED with improved linearity for 3 Gb/s free-space VLC

without pre-equalization, IEEE Xplore digital library. *IEEE Photonics Journal, 8*, 3. https://doi.org/10.1109/JPHOT.2016.2564927.

166. Chun, H., Rajbhandari, S., Faulkner, G., Tsonev, D., Xie, E., James, J., McKendry, D., Gu, E., Dawson, M. D., O'Brien, D. C., & Haas, H. (2016). LED based wavelength division multiplexed 10 Gb/s visible light communications. *Journal of Lightwave Technology, 34* (13), 3047–3052.

167. Chi, N., Zhang, M., Zhou, Y., & Zhao, J. (2016). 3.375-Gb/s RGB-LED based WDM visible light communication system employing PAM-8 modulation with phase shifted Manchester coding. *OSA Publishing Optics Express, 24*, 21663–21673.

168. Ferreira, R. X. G., et al. (2016). High bandwidth GaN-based micro-LEDs for multi-Gb/s visible light communications. *IEEE Photonics Technology, 23*(19), 2023–2026. https://doi.org/10.1109/LPT.2016.2581318.

169. Nasir, S., Shuaishuai, G., Ki-Hong, P., et al. (2019). Optical camera communications: Survey, use cases, challenges, and future trends. *Physical Communication Journal, 37*. https://doi.org/10.1016/j.phycom.2019.100900.

Chapter 2
Conventional and Advanced Technologies for Wireless Transmission in Underground Mine

2.1 Short Survey on Conventional Communication Systems in Underground Mine

Due to an avalanche of smart devices with wireless communication capability integrated, the well-known wireless communication technologies and applications based on RF (Wi-Fi, BLE, world interoperability for microwave access—WiMAX, Bluetooth, ZigBee, Z-Wave, 6LowPAN, RFID, Ultra Wideband—UWB, NFC) are increasingly used, lately. These wireless technologies and applications based on RF, intense exploited every day by a significant number of users, have become too crowded and almost overwhelmed, especially indoors, where the most wireless transmissions take place [1].

Some of the RF technologies have been already investigated and tested in underground mine, and a number of them are successfully applied in spaces where the environmental conditions allow the use the RF signals. For example, Bluetooth, Wi-Fi, UWB, WiMAX, RFID, or ZigBee technologies and applications for wireless transmission have been tested and, in some cases, deployed for underground communication, tracking, monitoring, and personnel positioning.

The first known experiment to detect radio signals into underground mines has been done in 1922 by US Bureau of Mines (USBM) in Pennsylvania [2]. The main objective of the research project of USBM was to test on the propagation of radio waves through the ground in order to track and fast rescue miners following an accident. Research results showed that electromagnetic EM signals at ultralow frequency (ULF) (from 630 Hz to 2 kHz) were transmitted through mine rock up to a distance of 1645 m. The prototype developed by USBM for the wireless communication system used state-of-the-art technology of that time with off-the-shelf components to reach targeted reliability and low cost. The innovative technology enabled a real-time warning of the underground personnel, by inducing flashing on their cap lamps [3].

© Springer Nature Switzerland AG 2021
S. M. Riurean et al., *Application of Visible Light Wireless Communication in Underground Mine*, https://doi.org/10.1007/978-3-030-61408-9_2

The EM detection and through the earth communication aiming to shorten the time till the miners' rescue after an underground accident or disaster, using narrow-band transmitter has been investigated, described, and developed by many authors. Most of the early developed systems aimed to permit signals to be continuously send to the surface from the point where the miners were trapped underground due to roof falling or collapse of side gallery [4–8].

Although these systems proved to be very useful, even in lives saving, there are limitations for their world wide application in underground mines, because the radio waves can be propagated to the targeted distance only if the environment has the appropriate physical and electrical properties [9, 10].

The most important property of the signal transmission, signal attenuation, depends mainly upon the physical properties of the environment such as obstacles in the propagation path, rock density, wall roughness, entry tilts, and so on [11]. Other effects that also influence attenuation are earth conductivity and dielectric constants. The low frequency of the EM field can penetrate kilometers of rock and soil; therefore, the signal will reach surface from deep underground making it suitable for different underground applications. Many investigations regarding the EM waves' propagation (600 Hz–60 MHz) through rock have been developed by researchers in different parts of the world [12, 13] and a reliable communication has been proved and settled (at 200 m depth) in coal mines in South Africa [14].

An interesting application of the EM wave modulation for early detection of miners' position (up to 90 m in LoS or 20 m through walls and debris) while they are in dangerous situations underground was developed in 1995 by *Selectronics* (Germany) and manufactured under the name SIRUS. The system relies both on the heartbeat and chest' movement while miners are breathing. A microwave transmitter with directional antenna and a Doppler receiver detects the reflected signals, both cardiac and breathing [7].

Thanks to researchers' efforts in universities and government agencies around the world, some useful commercially available products have been launched. A reliable system (consisting in dedicated coded belt-wearable miner's tag—with a buzzer and LED integrated—and a portable search unit) assigned to detect and locate trapped miners underground (at a distance of more than 30 m through rock) has been used at CSIR Mining Tek in South Africa [15].

In Austria, at Schwaz, a system consisting of a beacon enclosed into the miner's cap lamp and a handheld location receiver used to search for the trapped miner's beacon was successfully tested in a mine in Tirol with a detection accuracy of 50 cm [16].

There is an ever-increasing demand of a real-time surveillance of personnel, equipment, and different kind of data acquired by sensors to inspect the air quality, for example, in order to increase the underground safety and security at work.

More and more advanced technologies are applied and developed for underground, aiming to get a continuous monitoring of the mining entire activity.

Local underground wireless communication and remote data-transfer solutions have already been applied worldwide in underground mines since 2000 [17–22].

New, emerging technologies have also been researched and proposed to be implemented in mining industry, for example, the use of robots with remote wireless communication for real-time surveying or rescue operations [23–26], advanced wireless tracking technologies [27–29] based on Wi-Fi, RFID, ZigBee, Bluetooth, UWB, or hybrid systems [30–33], all of them aiming to improve, on one hand, the workers' safety and then productivity in underground mines, and on the other hand to provide fast and reliable rescue systems.

Although a hybrid communication consisting in wireless and cable-based (PLC, PoE, coaxial cable, twisted pairs, or optical fiber) underground to the surface transmission, seems to the most appropriate applicability in mining industry, there are many issues needed to be overcome related both to the RF wireless and cable-based communication.

The use of cable-based communication in underground mining industry has its own drawbacks [34], as follows:

- the entire communication system becomes vulnerable in case of emergency, due to cable breakage,
- risk to appear sparks/flames that would seriously threaten workers' security and safety underground,
- lack of efficient coverage, because new working areas are opened every day due to the methods of mining by advancement,
- support for only point-to-point communications,
- mobility of personnel and equipment is limited or seriously reduced.

The use of RF wireless communication has also its drawbacks since it mainly suffers from *signal attenuation* and experiences EMIs.

Signal attenuation occurs due to the environmental intrinsic properties of the transmission optical medium (absorption, reflection, diffraction, scattering, and bending losses) that can be filled with smoke, tiny suspended particles of rock and tiny drops of water in air, the obstacles (from different materials with lower or higher density) on the path of the signal communication and, of course, natural occurring (and linearly increasing) over a long-distance transmission.

The signals' power attenuation A_p in decibels (dBs) is given by the formula:

$$A_p = 10 \; log \;_{10} \frac{P_{Tx}}{P_{Rx}} \tag{2.1}$$

where:

P_{Tx}—the signal power at source (transmitter—Tx).
P_{Rx}—the signal power at destination (receiver—Rx).
And $P_{Tx} > P_{Rx}$.

Signal also experiences high attenuation when it encounters corners, therefore, propagation around multiple corners is severely attenuated. Although high frequencies are likely to be used in straight, open spaces (as main galleries underground are) they suffer from great loss when passing over a number of corners [35]. Therefore, the choice of signal's frequency depends on the space configuration.

EMI can seriously affect the operation of electronic devices when it is in the vicinity of an EM field in the RF spectrum that is caused by another electronic device. These interferences can decrease the performance of sensitive wireless receivers nearby.

Also, an important limiting factor in using RF communication system is that the machinery used in underground environments randomly creates a wide range of many types of intense EMIs during everyday mining operation.

In mines with potential risk of explosion, all equipment must be shielded to keep unwanted RF energy from entering or leaving. Depending on the level of risk the mine has, there are standards that rule the class of protection for any type of equipment underground, starting from illumination, air ventilation, transport equipment, machinery, and so on. Electromagnetic noise amplitude decreases with increase in frequency [36].

Therefore, alternative to RF wireless networks, new wireless data communication technologies are not only necessary today and in the near future, but compulsory.

2.2   Wireless Transmission Based on Visible Light. A Detailed Blueprint

In the optical spectrum, the IR together with visible light regions is about 2600 times larger than the RF spectrum (0–300 GHz) [37].

The general characterization of a VLC channel is done by its optical channel impulse response (CIR), which is used to investigate and decrease, as much as possible, the effects of channel noise. Both experimental measurements and computer modeling approach on the channel characterization of underground and outdoor systems have already been described. Figure 2.1 shows a general block diagram of a conventional VLC system.

The power penalties associated with the optical channel are usually separated into optical path loss and multipath dispersion that expresses itself as the intersymbol

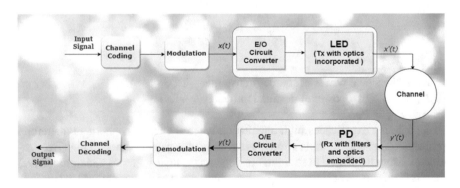

Fig. 2.1 General diagram of a VLC system

interference (ISI). Dispersion is modeled as a linear baseband channel impulse response (CIR) $h(t)$.

The channel characteristic of a VLC link is stable in case of:

- a given position of oTx,
- a given position of oRx,
- accurate reflecting characterization of the underground surroundings and objects.

The general optical channel model is given by formula:

$$y(t) = R \cdot x(t) \otimes h(t) + n(t) \tag{2.2}$$

where:

$\mathscr{R}$—the photodetector' (PD) responsivity.

$x(t)$—transmitted optical signal—since in optical wireless systems the instantaneous optical power is proportional to the generated electrical current, $x(t)$ represents the power signal.

$\otimes$—convolution.

$y(t)$—signal received at PD.

$h(t)$—optical CIR.

$n(t)$—total noise consisting of ambient noise in optical channel, shot noise, and thermal noise in oRx entire front-end device.

Ambient noise comes from natural and different other artificial light.

Shot noise refers to fluctuations in the number of photons sensed in PD according to their occurrence independent of each other. Shot noise is one of the main noise source in the OW link and arises fundamentally due to the discrete nature of energy and charge in the PD.

Thermal noise (also known as Johnson noise) arises in all conducting materials and is produced by the thermal variation of electrons in a receiver circuit of corresponding resistance and temperature (T_e) [38]. The electrons are in constant motion, and they frequently bump into the molecules or atoms of the substance. Every free flight of an electron creates a minute current. The sum of all these currents during a long period of time is equal to zero. This is because the power spectral density (PSD) does not depend on frequency. Furthermore, the AWGN follows the Gaussian distribution with zero mean and variance for IM/DD and coherent receivers.

As shown in Fig. 2.1, E/O stands for converting electrical signals to optical signals, and O/E converts optical signals into electrical signals. In the frequency domain, the channel can be characterized by its frequency response, which is the Fourier transform of the impulse response.

In order to evaluate the overall performance of a particular type of communication topology, it is important to know the distribution of the channel characteristics over the entire area and to take into consideration the worst receiver position, since that describes best the lower limit for the link performance [39].

Considering the IR communication, the optical channel impulse response (CIR) ($h(t)$, Eq. 2.3), was for the first time modeled by *Gfeller* and *Bapst* [40], as follows:

$$h(t) = \begin{cases} \dfrac{2t_0}{t^3 \sin^2(FoV)}, & t_0 \le t \le \dfrac{t_0}{\cos(FoV)} \\ 0, & elsewhere \end{cases} \qquad (2.3)$$

where:

FoV—field of view.

t_0—the minimum delay.

The optical signal transmitted has the following constraints:

- $x(t)$ is real and positive $x(t) \ge 0$,
- the maximum optical transmit power is limited by the eye safety photobiological standards and regulations EN 62471–2006 and 60,825–2008.

Therefore, the average value of $x(t)$ should not exceed a specified maximum power value P_{max}, that is:

$$P_{max} = \lim_{T \to \infty} \frac{1}{2T} \int_{-T}^{T} x(t)dt \qquad (2.4)$$

In VLC systems, high optical transmit power is required, therefore only a small path loss of the optical signal is allowed. In this regard, the most suitable modulation techniques are high peak-to-mean power ratios that can be achieved by trading-off the bandwidth against the power efficiency. The signal-to-noise ratio, in case that the shot noise is dominant, is proportional to the active area of the PD (A_{PD}^2).

The PD's capacitance increases proportional with its active area, resulting in a limited bandwidth received by PD and, therefore, the transmission capacity is restricted.

So, the two main trade-offs in optical communication consist of increased bandwidth requirement connected with power-efficient modulation techniques.

The $H_{VLC}(f)$ of the VLC is:

$$H_{VLC}(f) = H_{LoS} + H_{dif}(f) \qquad (2.5)$$

where:

H_{LOS}—the transfer function in line of sight (LoS) scenario (dependent on the distance and objects set between oTx and oRx and their orientation but independent on the modulation frequency).

H_{dif}—the transfer function in environments with diffused link.

In most of the research papers, the transfer function in environments with diffused link (H_{dif}) has not been taken into consideration because most indoor environments are homogeneous and isotropic. The underground mine environment or industrial halls are not considered here and not yet deeply investigated, as far as the authors know.

Kahn and *Barry* characterized for the first time the indoor optical wireless channel for communication (for IR spectrum) in these two scenarios: line of sight (LoS) and non-line of sight (NLoS), based on the relative strength of light (signal components) between the LoS and NLoS [41]. The path of propagation of oTx and detected by oRx characterize the two link arrangements in LoS and NLoS communication.

Regarding the geometry of a VLC scenario, there are important underground position of oTx and oRx, their radiation/detection characteristics as well as the reflection properties of the objects in the specific environment under study.

In order to achieve the best communication between oTx and oRx, depending on space's geometry, the number and type of objects into the studied area (walls, ceiling and floor, furniture, including windows and doors) and their radiation/detection characteristics, the network has to be positioned as to maximize both the coverage and optical capacity over the room where the VLC is setup [39].

However, the VLC network setup has to take into account that the primary functionality of the network is the underground's illumination. The spotlighting at oRx can improve the irradiation intensity of the oTx [42].

Many studies have already been made on VLC channel and spatial conditions related to the optical network configuration based both on direct measurements [43] and ray-tracing simulations [44].

The main building blocks (general structure) of a transmission link in VLC are presented in Fig. 2.2.

A digital signal processor (DSP) together with digital to analog converter (DAC) is used generally in VLC setups in order to apply the necessary modulation technique with the aim to code and convert the information necessary to be wireless remotely sent through the optical channel.

The electrical signal (current) drives the LED (or an array of LEDs) where the information-carrying current signal is transformed into optical intensity.

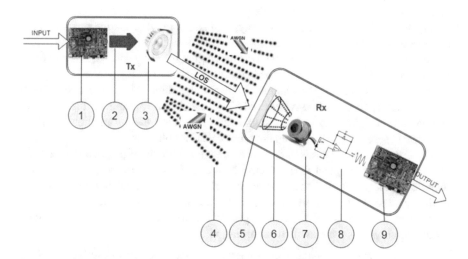

Fig. 2.2 Main blocks of a transmission link in VLC (Adapted from [39]). 1—DSP + DAC, 2—signal, 3—LED with optics, 4—optical wireless channel (white light), 5—Filter, 6—Lens (optical concentrator), 7—PD, 8—TIA, 9—DSP Board

When between the oTx and oRx a high distance is planned to be reached, different optical components are used, positioned in front of the LDs (laser diodes) or μLEDs (consisting of one or many chips arranged in series or parallel in order to act as a single emitter) in order to achieve higher output power. The optical components consist of lenses of different shapes and dimensions, collimators, or diffusers that shape the transmitted beam in such a way that is going to be optimum for the specific scenario planned.

The optical signal is now transmitted through space, named *optical wireless channel*. The PD collects both the desired optical signals and other lighting signals known as AWGN with a negative effect on the VLC communication quality. In front of the PD is usually used as an optical filter with the aim to decrease the effect of AWGN produced by ambient light. The original optical signal's quality decreases is attenuated both due to the optical path length and to the first major component of the AWGN, the background illumination, and the sunlight (which is not applicable underground). An optical non-imaging concentrator is also used in order to improve the signal-to-noise ratio (SNR) of the PD.

Some amount of the optical energy is also absorbed by different objects (ceiling, floor, furniture, etc.) in the environment, and some are reflected back in a diffuse or specular way. Both LoS and NLOS signals arrive at the oRx with different delays that depend on a number of environmental characteristics.

In order to select the necessary part from the optical spectrum, to avoid the slow response of the yellow color emitted by the white LED, to reduce the interference from ambient light, a blue wavelength optical filter is used in front of the oRx.

Afterward, the optical signal passes through a system of optical elements. These optical elements (collimator lenses) have the role to amplify the signal, align the light beam, and spot it on the active area of the PD for an optimum detection of the original optical signal sent by the oTx.

The PD (a PIN PD, an APD, or arrays of PDs) converts back the optical signal into electrical current. An electrical low-pass filter is used to filter all the undesirable noise components. Next, the received current signal is very weak, so, it is electronically pre-amplified with a transimpedance amplifier (TIA).

In order to recover the original data transmitted, a digital signal processor (DSP) with an analog to digital converter (ADC) is used, transforming in this way, the analog current signal into a digital signal and demodulate the information bits [39].

When is necessary to decide on the proper VLC topology in specific environment for underground applications, a correct estimation of the light propagation, characteristics of both LEDs and PDs, the environment particularities, the space geometry including reflections, and the channel model with the channel impulse response (CIR), have to be all taken into consideration and deeply investigated.

Two of the most used light propagation topologies for VLC's indoor/underground applications are classified according to:

– the degree of directionality between oTx and oRx (e.g., direct LoS as shown in Fig. 2.3),
– direct or diffuse LoS or NLoS light's path between oTx and oRx (Fig. 2.4).

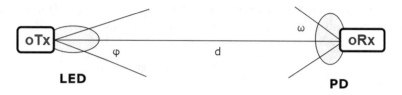

Fig. 2.3 Topology of a direct LOS communication between oTx and oRx

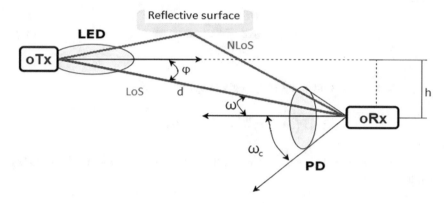

Fig. 2.4 Geometrical model of a LoS VLC link

A direct LoS link provides at the oRx a high irradiation intensity and a wide channel coherence bandwidth (Fig. 2.3). Where high user mobility is required, LoS is not a suitable setup, since the link can be easily interrupted or blocked by other users.

In the direct LoS link, oTx and oRx are straight aligned. Although the advantage of this topology is high power utilization, it requires alignment of both terminals, because once there is an obstacle in the transmission path, the data communication will be obstructed. This is the best choice topology for a point-to-point communication suitable for two, close to each other, static terminals as long as there are no obstructive objects between oTx and oRx.

Regarding the diffuse LoS topology for data communication, in order to avoid the system to be more or less affected by the shadow effect, the directional requirements of oTx and oRx must be reduced. This allows the oRx to use a larger perspective to realize an enhanced communication between oTx and oRx. The optical power will be, in this case, uniformly distributed into the room, but, on the other side, the multipath effects of the link limit the signal's transmission rate.

In case of a shadowed LoS channel, the model is quite difficult to be described, since the LoS path is blocked. The entire optical CIR is determined by reflections which, as for LoS unshadowed channels, are quite difficult to predict. Hence, neither the path loss nor the large delay spread is easily predicted.

For the generalized Lambertian radiant intensity, the angular dispersal of the radiation intensity pattern has the following distribution [45]:

$$R_0(\varphi) = \begin{cases} \dfrac{(m_1 + 1)}{2\pi} \cos^{m_1}(\varphi), & \varphi \in \left[-\dfrac{\pi}{2}, \dfrac{\pi}{2}\right] \\ 0, & \varphi \geq \dfrac{\pi}{2} \end{cases} \tag{2.6}$$

where:

$$m_1 = \frac{-\ln 2}{\ln\left(\cos \varphi_{\frac{1}{2}}\right)} \tag{2.7}$$

The radiant power intensity $P(\varphi)$ is given by:

$$P(\varphi) = P_t \frac{m_1 + 1}{2\pi} \cos^{m_1}(\varphi) \tag{2.8}$$

The active area of PD, A_{eff_PD} collects the radiation incident at angle ω smaller than the PD's FoV:

$$A_{eff\_PD}(\omega) = \begin{cases} A_{PD} \cos \omega, & 0 \leq \omega \leq \dfrac{\pi}{2} \\ 0, & \omega > \dfrac{\pi}{2} \end{cases} \tag{2.9}$$

A large-area PD seems to be suitable for indoor/underground VLC in order to collect as much power as possible but in reality, this situation draws a lot of additional problems:

- increased junction capacitance and therefore decreased receiver bandwidth,
- increased manufacture cost,
- increased receiver noise.

There is, however, a solution for this situation that stands in the use of a non-imaging concentrator in front of PD in order to increase the overall effective collection area of the oRx.

An ideal non-imaging concentrator has an *optical gain* $g(\omega)$ represented by the expression:

$$g(\omega) = \begin{cases} \dfrac{n^2}{\sin^2 \omega_c}, & 0 \leq \omega \leq \omega_c \\ 0, & \omega > \omega_c \end{cases} \tag{2.10}$$

where:
n—internal refractive index.

$\omega_c \leq \pi/2$ is the FoV.

The FoV of the oRx is related to the of PD's lens $A_{col\ PD}$ and the PD active area [16, 40] as:

$$A_{col\ PD}\left(\frac{\omega_c}{2}\right) \leq A_{PD} \tag{2.11}$$

Results that, there is an inverse proportional relation between the FoV and the concentrator gain: when the FoV is reduced, concentrator gain increases. When the length between oTx and oRx is of the order of few meters or less, the optical power attenuation due to the absorption and scattering is low.

The DC gain of oRx in an LoS scenario for VLC system, with a Lambertian source having $m_1 = 1$, a band-pass filter, and a non-imaging concentrator is [46]:

$$H_{LOS} = \begin{cases} \dfrac{A_{PD}(m_1 + 1)}{2\pi d^2} \cos^{m_1}(\varphi)T_s(\omega)g(\omega)\cos\omega, & 0 \leq \omega \leq \omega_c \\ 0, & \text{elsewhere} \end{cases} \tag{2.12}$$

where:

$T_s(\omega)$—optical transmission of the band-pass filter.
d—distance between oTx and oRx (see Fig. 2.3).

The optical intensity received becomes in this case:

$$P_{PD\_LOS} = H_{LOS}(0)P_t \tag{2.13}$$

In case of an indoor/underground VLC with short distance oTx and oRx, the multipath dispersion is not taken into consideration, a LoS link channel is modeled as a linear attenuation and delay. Therefore, the system is considered as non-frequency selective and the path loss depends on the square of distance between the transmitter and the receiver.

In this case that $\varphi < 90°$, $\omega <$ FoV, and $d \gg \sqrt{A_{PD}}$, the optical CIR is given by:

$$h_{LoS} = \frac{A_{PD}(m_1 + 1)}{2\pi d^2} \cos(\varphi)T_s(\omega)g(\omega)\cos(\omega)\delta\left(t - \frac{d}{c}\right) \tag{2.14}$$

where:

c—speed of the light in free space (3×10^8 m/s).
$\delta(.)$—Dirac function.
$\delta(t - d/c)$—signal propagation delay.

An NLoS VLC propagation scenario is generally known as diffuse system that takes into consideration reflections off the room's surfaces (ceiling, walls, floor, window, door, and their reflection's index due to specific surface composition) as well as furniture or any other type of static or dynamic obstacles inside the room.

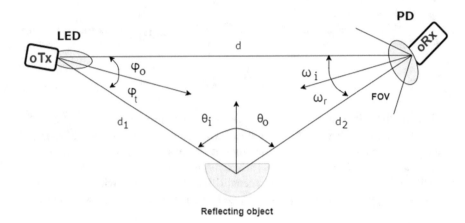

Fig. 2.5 Geometrical model of a directed NLoS with a single reflection LoS light propagation topology

NLoS is generally described as a light propagation scenario that takes into account the signal arriving from oTx to oRx after one or multiple reflections or bounces from the objects inside a room, being separately considered by the directivity of the oTx relative to oRx. However, there are two different NLoS propagation scenarios:

- directed NLoS,
- diffuse NLoS.

In case of *directed* NLoS, the oTx has a narrow radiation characteristic, projecting a single spot of light on the ceiling, for example. The ceiling, in this case, acts as a new oTx, conveying the strong light intensity to the oRx based on its reflection characteristics.

As for the case of *diffuse* NLoS, oTx has a wide radiation characteristic irradiating a large portion of the reflecting surface. The light will arrive at the oRx after one or multiple reflections (bounces) on the different surfaces of objects indoor/underground.

In some circumstances, these reflections are considered as unwanted signals since they carry out multipath propagation and thus distortions, ray tracing delay, and therefore, the estimation of the path loss has a potential growth.

In Fig. 2.5, the correlated orientation of oTx and oRx as well as orientation toward the reflecting object is described highlighting both the observation angles θ_o and incident angles θ_i.

Low path loss is experienced with *high reflective objects* underground, due to the increased power of the rays coming from different path of propagation. This situation drives to low channel bandwidth and high delay spread (Fig. 2.6).

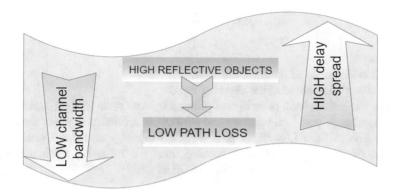

Fig. 2.6 Delay spread and coherence bandwidth for high reflective objects

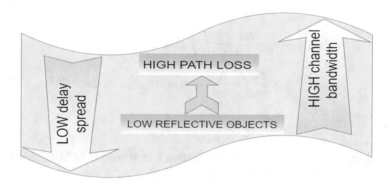

Fig. 2.7 Delay spread and coherence bandwidth for low reflective objects

On the other side, when objects underground are *low reflective*, high path loss is expected, leading to higher channel bandwidth and consequently a lower delay spread (Fig. 2.7).

In case of underground mining where environments are highly polluted and in some industrial halls, characteristics of the propagating optical beam are highly affected, thus resulting in optical losses and turbulence-induced phase fluctuation and amplitude fades on the scale of a wavelength.

Gfeller and Bapst [40], the first researchers who modeled the path loss of an IR wireless channel, presented an analytical model of the received optical power in IR radiation, both for LoS and one reflection NLoS. They also took into consideration in their study, the orientation of oTx and oRx as well as orientation in the direction of the reflecting surface.

Most of the researchers so far took into consideration the VLC model with the oTx modeled according to a Lambertian radiation pattern and the oRx according to a Lambertian detection pattern with a specific FoV.

The bidirectional reflectance distribution function (BRDF) $R(\theta_i)$ is:

$$R(\theta_i) = \frac{\cos \theta_o}{\pi} \qquad (2.15)$$

where:

θ_i—incident angle of the incoming light.

θ_o—observation angle of the outgoing light.

Thus, the general form of an optical power received on the PD (P_{PD}) active area, in a diffuse single path LoS topology and a BRDF, is defined as a sum of the optical power received from LoS direct path (P_{LoS_dp}) topology and the optical power received from a single reflection coming from the reflecting surface (object) (P_{NLoS_sr}), taking into consideration the BRDF as well:

$$P_{PD} = P_{Los\_dp} + P_{NLoS\_sr} \qquad (2.16)$$

$$P_{LoS_{dp}} = P_{LED} \frac{m_1 + 1}{2\pi} \cos^{m_1}(\varphi_o) \frac{A_{PD}}{d^2} \cos(\omega_i) r_{FOV}(\omega_i) \qquad (2.17)$$

$$P_{NLoS_{sr}} = P_{LED} \frac{m_1 + 1}{2\pi} \int_\varphi \int_\omega \cos^{m_1}(\varphi_t)$$
$$\times \frac{\rho R(\theta_i)}{d_1^2} \frac{A_{PD}}{d_2^2} \cos(\omega_r) r_{FOV}(\varphi_t) d\varphi d\omega \qquad (2.18)$$

where:

P_{LED}—total optical power of LED.

m_1—Lambertian number of LED's radiation lobe related to LED directivity.

φ_o—observation angle of oTx on the direct path.

ω_i—incident angle of the oRx.

A_{PD}—active area of the PD.

d—distance between oTx and oRx on the direct path

$$r_{FoV}(\omega_i) = \begin{cases} 1 \ for \ \omega_i \le \omega_r \\ 0 \ for \ \omega_i > \omega_r \end{cases} \qquad (2.19)$$

ρ—reflection coefficient of the reflecting object according to its material's surface.

d_1—distance between oTx and the reflecting object.

d_2—distance between the reflecting object and oRx.

However, in a real scenario, the light behavior is much more complex and the VLC setup has to take into consideration the oTx emittance spectrum, oRx spectral response in a function of wavelength (λ), as well as the underground objects and their geometry and the type of materials in the surrounding environment.

Still, in a VLC setup, WLED has a wide range of emittance spectrum and the spectral response of the oRx is close to the spectral response of the oTx, therefore P_{PD} (Eq. 2.16) depends on the wavelength (λ). When a complex underground geometry and material characteristics of the objects impose multiple reflections, light undergoes multiple reflections, and a ray-tracing algorithm should be applied [39].

As *Pakravan* et al. showed in the paper "*Indoor Wireless Infrared Channel Characterization by Measurements*" [47], for an oRx changing the elevation angle from $0°-180°$, not more than five points along the entire path are enough to calculate variation of channel path loss for a full rotation range, therefore, the statistical path loss in a VLC setup is a simple curve fitting that can be approximated with intermediate values.

We have already underlined that the optical signal experiences multipath fading if PD size (i.e., the active surface area) is proportional to one wavelength or less. Fortunately, VLC receivers use PDs with a surface area usually millions of square wavelengths.

Furthermore, the total photocurrent generated is proportional to the integral of the optical power over the entire PD's active area.

2.2.1   Optical Setup

Optical Transmitter

LEDs are used as light sources in VLC systems aiming to piggyback data transmission along with illumination. The optical signal transmitted by LEDs experiences, invariably, free-space diffusion to the receiver, which is usually one or more photodetectors (PDs).

A LED consists of a number of semiconductor materials overlaid. Inside the LED, electricity is converted directly into light particles, photons, leading to an efficient result (gain) related to other light sources where only a slight amount of the electricity is converted into light and for the most part into heat. Electric current is used both in incandescent bulbs (IB) and halogen lamps (HL) to heat a wire filament, making it glow. In case of fluorescent lamps (FL), it is produced a gas discharge that creates both heat and light. On the other hand, LEDs, compared to classical light sources, need a low level of energy to emit light. Furthermore, LEDs are improved every day, getting more efficient with higher luminous flux (measured in lumen [lm]) per unit electrical input power (measured in watt [W]).

The main "actor" of the optical transmitter (oTx), the optical signal emitter in a VLC setup, can be any of the following:

- white LED,
- RGB LED.
- an array (attocells or picocells) of μLEDs,
- laser diode LD.

Because there are many advantages of using the LEDs in underground spaces for the illumination networks, for the miner's cap lamp (placed on the helmet) and the machinery spot lights, the solid-state lighting (SSL) is expected to replace the existing lighting infrastructure in the underground mines. Some of the most important advantages of SSLs are summarized as follows:

- long lifetime,
- high energy conversion efficiency,
- minimum heat generation characteristics compared to all other lighting sources,
- high tolerance to humidity and extreme (high/low) temperatures,
- mercury free,
- small with compact size,
- most important, allow fast switching,
- can operate for 25,000–50,000 h before their output drops; LEDs are more long lasting than other lamps, hence decreasing materials consumption. IB tends to last 1000 h, as heat damages the filament, while FL typically last around 10,000 h,
- 1/4 of worldwide electricity consumption is used for lighting, therefore, the high energy/efficient LEDs can contribute to saving the Earth's resources. This opportunity opens a bright road for the various applications of wireless optical communication.

However, despite all the advantages underlined above, there are still a number of drawbacks and technological challenges needed to be addressed by researchers worldwide in various fields, from electronics to lighting, communication, and so on. Some of the most important challenges refer to the designing of low cost, high efficient front end devices with high luminous efficiency, and outstanding color quality since the primary function of LEDs still should be illumination.

All the listed advantages and opportunities due to advanced technologies developed today, give the possibility to efficiently aggregate lighting, along with data wireless communication embedded into different network types:

- Miner to miner (M2M).
- Miner to illumination–infrastructure/illumination–infrastructure to miner (M2I/I2M).
- Mining equipment to miner/miner to mining equipment (E2M/M2E).
- Mining equipment to illumination–infrastructure/illumination–infrastructure to mining equipment (E2I/I2E), and so on.

So, a special design for LEDs is requested with innovative optical solutions embedded in order to bring reliable glare-free along with energy-efficient solutions for every workspace underground as well as high-quality products that solve the challenges end-users encounter. Instead of covering the light sources, or to tilting the work light to avoid miners blinding each other, or in the vehicles' proximity, different companies found ingenious optical solutions that solve this problem [48]. These solutions still allow a reliable optical data communication along with illumination.

Efficient LEDs have more than 300 lm/W, which can be compared to 16 for regular IBs and about 70 for FLs.

The power (P) of a LED in watts [W] is equal to the luminous flux (Φ_V) in lumens [lm], divided by the luminous efficacy η in lumens per watt [lm/W]:

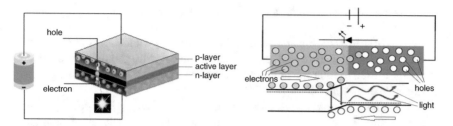

Fig. 2.8 The inside structure of a LED

$$P_{(W)} = \frac{\varnothing_{v(lm)}}{\eta_{\left(\frac{lm}{W}\right)}} \qquad (2.20)$$

A LED consists of several layers: *p-type layer* with a deficient amount of electrons (also called surplus of positive holes), *n-type layer* with an extra number of negative electron and a special layer between them named *active layer* (Fig. 2.8). When an electric voltage is applied to the semiconductor, both the positive holes and the negative electrons are driven between these two layers into the active layer. The light is created at the moment when electrons and holes meet.

The LED itself is no bigger than a grain of sand and the light's wavelength depends on the semiconductor used during the manufacturing process. For example, blue light has the wavelength between approximately 380 and 500 nm, making it one of the highest energetic efficient with the shortest wavelengths that can be produced by certain materials.

> "Creating light in a semiconductor LED technology originates in the same art of engineering that gave us mobile phones, computers and all modern electronics equipment based on quantum phenomena" [49].

Henry J. Round and *Guglielmo Marconi* reported for the first time, in 1907, a discovery based on a semiconductor that emitted light and therefore they became Nobel Prize Laureates in 1909 [50].

Oleg V. Losev in the 1920s and 1930s, in the Soviet Union, carried out closer studies of light emission [51]. However, both *Round* and *Losev* didn't reach a deep understanding of the phenomenon discovered.

Few decades later, the prerequisite for a theoretical description of the electroluminescence was created. Although at the end of the 1950s, was invented the red light-emitting diode, a blue diode—was actually needed to create white light [52].

Researchers fought to find out how to set the appropriate angle with which the light escapes the semiconductor, since this "light cone" is quite narrow. Following many attempts, they figured out how to intensify the light output by forcing light to refract and bounce off all surfaces of the crystal's semiconductor. This is the reason why traditional LED displays have been properly viewed only from one best angle.

The light-emitting diode in this type of lamp that consists of some diverse layers of gallium nitride (GaN). By mixing in indium (In) and aluminium (Al), the Nobel

Prize laureates (*Akasaki*, *Amano*, and *Nakamura*) succeeded in increasing the lamp's efficiency.

Red and IR LEDs are made with gallium arsenide, white LEDs are made with yttrium aluminium garnet and bright blue is made with gallium nitride (GaN). Phosphor is used to filter the light that goes out of the LED, therefore, being created a cleaner "harsh" white color [53, 54].

In 1986, both *Akasaki* and *Amano* were the first to produce a high-quality gallium nitride crystal. They positioned over a sapphire substrate, a layer of aluminium nitride and then, on top of it, placed a high-quality gallium nitride. At the end of the 1980s, *Akasaki* and *Amano* succeeded and create a *p-type* layer. It was a chance for *Akasaki* and *Amano* to notice that, when it was studied in a scanning electron microscope, the material used was glowing more intensely and therefore the *p-type* layer made was more efficient under the electronic beam of the microscope. Finally, in 1992, they succeeded to present the first diode emitting a bright light of blue color.

In 1988, on the other side of the world, *Nakamura* began to develop the blue LED and in 1990 he also succeeded in creating high-quality GaN LED. He created the crystal in his own different way, by growing first a thin layer of GaN at low temperature, and then growing successive layers at a higher temperature.

Both research groups succeeded, during the 1990s, to improve the blue LEDs, making them more efficient. They made different GaN alloys by using indium or aluminium, therefore, the structure of the LEDs became more and more complex.

Amano, together with *Akasaki*, as well as *Nakamura*, invented also a blue laser, in which, the blue LED, is an essential component. A blue laser emits a cutting-sharp beam, in contrast to the dispersed light of the LED. As for the disk digital data storage supports, since the light colored blue has a very short wavelength, four times more information can be stored by the same area compared with infrared light. Blu-ray discs were therefore developed having increased storage capacity and longer playback times, as well as better laser printers.

Today, there are basically two types of LEDs as energy-efficient emitters of white light: a single-color LED (obtained from the tree colors red, green, or blue) and a white LED (WLED). These two technologies generate white light via LEDs. One is by combining RGB (red ~625 nm, green 525 nm, and blue 470 nm) in an accurate quantity in order to generate white light. Usually, these three devices stand by one with three emitters and suitable optics, being often used in applications where adjustable color emission is required. These devices allow WDM.

The phosphorescent white LEDs are created by a different technique. This technique involves the use of GaN blue LED coated with an inorganic phosphor layer that emits yellow light. The layer of phosphor absorbs a part of a short wavelength light produced by the blue LED and then the radiated light from the absorber experiences wavelength change to a longer wavelength of yellow light. The red-shifted emission mixes additively with the non-absorbed blue part to create the required white color.

However, the slow response of the phosphor layer has an important disadvantage since it limits the modulation bandwidth of the phosphorescent white LEDs to just few megahertz [55].

In spite of the many advantages that WLEDs have compared with incandescent bulbs, halogen, or fluorescent lamps, their modulation bandwidth far exceeds that of the mentioned traditional lighting sources. The large area of WLEDs induces a large capacitance and slow response of the yellow phosphor that limit the 3 dB bandwidth to a few megahertz [56]. However, by removing the signal from the yellow phosphor that arrives at the oRx active area (with a filter), the 3 dB modulation bandwidth is possible to reach 10 or even more megahertz [57]. There are some references to different equalization circuit or digital signal processing used in order to extend the 3 dB bandwidth as well as a high order of multi-carrier modulation embedded for high spectral efficiency [58].

RGB mixed white LEDs provide a high spectral bandwidth, being more complex (the modulation circuit) and more expensive than W LEDs, therefore are not widely used in the VLC systems.

The emission of light is due to an electron' transition from an excited to a lower energy state. The energy difference leads to a radiative process that results in light generation. The recombination of the carrier is therefore used to provide flux photons.

Both the wavelength and frequency of the emitted/absorbed photons are related to the energy difference E of two energetic states (E_2 and E_1), and is given by equation:

$$E = E_2 - E_1 = hf = \frac{hc}{\lambda} \tag{2.21}$$

where:

h—Planck's constant (6.626×10^{-34} Js).
f—frequency.
c—speed of light (3×10^8 m/s).
λ—wavelength of the absorbed/emitted light.

Depending on the energy band gap of the semiconductor material, the radiated photons, can be in any part of the electromagnetic spectrum: UV, visible, or IR. In the case of LEDs, this conversion process is quite efficient. Therefore, a very slow amount of heat results compared to incandescent lights, for example.

The luminous efficiency of a LED is defined by the ratio of the luminous flux (in lumens) to the input electrical power (in Watt) (see Eq. 2.20). The luminous flux describes the "quantity" of total emitted light. The luminous flux (Φ_V) is the total emitted flux scaled properly to reflect the flexible sensitivity of the human eye to different wavelengths of light.

$$\phi_V = 683 \int V(\lambda) P_{sd}(\lambda) d\lambda \tag{2.22}$$

where:

683 [lm/W]—a normalization factor.
$V(\lambda)$—relative eye sensitivity (is normalized to unity at the peak wavelength of 555 nm).

$P_{sd}(\lambda)$—radiation power spectrum of the LED.

The optical power (P_o) emitted by a light source is then given by:

$$P_o = \int_\lambda P_{sd}(\lambda)d\lambda \qquad (2.23)$$

The off-the-shelf high-performance single-chip visible spectrum LEDs can have a Φ_V of about 10–100 lm at an injection current between 100 and 1000 mA.

A typical VLC link uses a white LED where both lighting and communication link are provided by the LED. The blue light can be extracted from the optical beam at the PD, using an optical filter.

The small-signal modulation bandwidth of the white LED depends on the LED driving current.

The most important characteristics of the oTx front-end device refer to:

(a) the radiation pattern,
(b) optical spectral response,
(c) electrical modulation bandwidth,
(d) electrical to optical (E/O) conversion.

The Radiation Pattern

The radiation pattern of a single LED is modeled by means of a generalized Lambertian radiation pattern shown in Fig. 2.9 [39].

The FoV of an LED is defined as the half-angle between the points on the radiation pattern $\varphi_{FoV,oTx}$.

An incoherent diffuse LED can have an FoV in the range between $\pm 10°$ and $\pm 60°$.

LED's Optical Spectral Response

In Table 2.1 [59] there are shown few common engineering terms used for specific photometric and radiometric quantities, with their corresponding SI units.

The oTx has to be manufactured according to BS EN 62471:2008 standard for photobiological safety of lamp systems and lamps. LEDs with incoherent diffusion, continuous wave modulated wave is part of the excepted group classification, and are not dangerous for human eye in case that, at a distance of 0.2 m, the irradiance is not higher than 100 W/m^2 from the optical source in the direction of maximal directivity within 1000 s [39]. However, the intensity can be substantially increased due to specific optical system of lenses and collimators and this situation has to be carefully conducted in case of a VLC or LiFi system that is constantly used indoor/ underground.

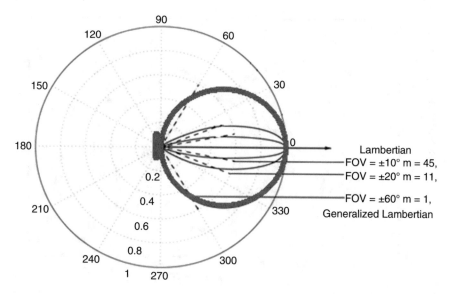

Fig. 2.9 Normalized Lambertian characteristics (Lambertian mode numbers, for $\pm 10° \pm 20°$ and $\pm 60°$ patterns and corresponding FOV. (Adapted from [39])

Table 2.1 Specific photometric and radiometric quantities

	Photometric (SI)	Radiometric (SI)
Flux	Luminous flux [lm]	Radiant flux [W]
Flux/area	Illuminance [lux = lm/m^2]	Irradiance [W/m^2]
Flux	Luminous intensity [candela = lm/sr]	Radiant intensity [W/sr]
Flux/(area/solid angle)	Luminance [nit = lm/(m$^2 \cdot$ sr)]	Radiance [W/(m$^2 \cdot$ sr)]

Today, on the market there are lots of incoherent diffuse IR LEDs with different optical centers of frequencies (830, 850, 870, 890, 940, 950) nm with a 3 dB bandwidth of approximately 40 nm. White LEDs have a spectral emission in the range between 380 and 780 nm (see Fig. 2.10).

LEDs Modulation Bandwidth

A VLC modulation bandwidth is directly influenced by the LEDs' frequency response, therefore, affects the rates of data transferred. The LEDs' modulation bandwidth depends on the lifetime (τ) of the carrier recombination on the active region and the *p–n junction* capacitance. Increasing the area of LED, its internal junction capacitance increases and consequently LED's RC delay time ($\tau = RC$) is also increased. RC delay time is defined as the necessary time for the voltage to rise from zero to about 63.2% of the DC voltage applied. Therefore, the modulation bandwidth of LEDs is restricted by great RC delay time. The frequency response of

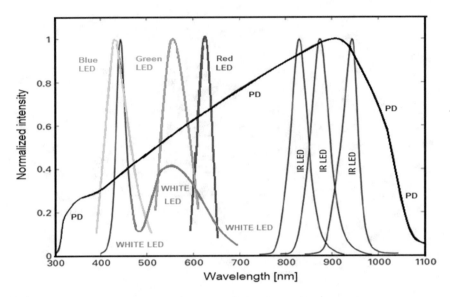

Fig. 2.10 LEDs and PDs optical spectral response in OWC (adapted from [39])

on-the-shelf white LEDs is highly influenced by the slow response of the yellow phosphor layer.

The modulation bandwidth of a VLC system is still limited to megahertz which is not good enough when trying to achieve data communication with high data rates. One of the solution explored is the use of a blue filter at the oRx to clean out the slow response of yellow phosphor.

Both the LED's frequency response and modulation bandwidth are related to the level of input current and the junction and parasitic capacitance.

The value of the capacitance is almost invariant, however, the response grows with increasing current, therefore, the effects of the above-mentioned factor can be decreased by superimposing the AC signal on a constant DC bias.

Considering the DC power (P) at frequency (ω), the relative optical power output at any given frequency is given by:

$$\frac{P(\omega)}{Po} = \frac{1}{\sqrt{1 + (\omega\tau)^2}} \tag{2.24}$$

The LED's cutoff frequency is also important. It refers to the maximum frequency at which LED's light emission drops to half of the initial light intensity.

Electrical to Optical (E/O) Conversion

LEDs, being complex semiconductors, are designed to convert an electrical current (electrical signal) into light (optical signal). The conversion process is fairly efficient in that it generates little heat compared to the incandescent bulbs or fluorescent tubes.

The LED's output power is linearly proportional to the drive current. As the LED is driven with higher currents, its chip gets hotter causing a drop in E/O conversion efficiency.

OLEDs

Organic light-emitting diodes (OLEDs) are SSL sources made of organic semiconductor (OSC) materials. SSL products, during illumination process, produce less heat and less energy dissipation due to the property of converting blue light in white light using photoluminescence, the same principle that is used for traditional fluorescent tubes [60].

Laser Diodes—LDs

An LD has the same basic structure as an LED in the sense that it contains p-type and n-type semiconductors, though, LDs are made with an additional region between the p–n junction, that has an intrinsic in nature, without dopants. Laser diodes have a threshold current which must be reached before lasing can occur. Before the threshold, LDs emit spontaneous light similar to LED. When the threshold current is reached the optical gain is exceeding the optical losses for the cavity. The cavity of an LD is created by the edges of the semiconductor, usually polished and cleaved to create a highly reflective side (99%) and a completely reflective side. This cavity delivers an oscillator for the emitted photons to become mobile back and forth. When photons are injected and transporting through the cavity, a photon stimulates an electron and hole recombination. The result is the emission of a photon duplicate to the one that produced the recombination. When this process occurs, and the amount of emitted light has become greater than the amount of absorbed light inside the cavity, lasing occurs.

The most advanced LDs feature the highest luminance of any light source available off-the-shelf today, with more than 1000 Mcd/m^2, which is more than 100 times higher than the brightest LEDs. The extraordinary illumination of LD combines the benefits of solid-state illumination, such as compact form factor, minimal power consumption, and long lifetime, with the highly directional output that has been possible only with legacy lighting technology [61].

The use of LD for different projects aiming to use LDs as efficient emitters for different setups of VLC systems has been reported in many research works worldwide with promising results [62–66]. In fact, LEDs and LDs are very similar devices and, when operating below their threshold current, all LDs act as LEDs.

Optical Receiver

The main "actor" of the optical receiver is the photodetector (PD) which is, in fact, a semiconductor with a *p–n junction*.

In a VLC setup, the PD can have any of the following:

- positive intrinsic negative photodetector (PIN-PD),
- avalanche photodiode (APD),
- an array of PIN-PDs or APDs,
- Light-dependent resistor (LDR).

The photons that arrive on the active area of the PD's surface, hit the p–n junction, and therefore, excite the electrons resulting in a current. A p–n junction diode is made when a p-type semiconductor is fused to an n-type semiconductor generating a possible barrier voltage across the diode's junction (see Fig. 2.11).

A PD can be operated in a (a) *photovoltaic* (zero biased) or (b) *photoconductive* (forward or reverse biased) mode.

In zero biased (photovoltaic mode), the cathode (C) and the anode (A) are connected to a load (the input of a TIA that converts photocurrent into voltage); therefore, the PD delivers a current (up to several mA) being used as sensor. PDs zero biased are used in power measurement applications.

When the PD is in the photoconductive mode (generally used for VLC), the incident optical power has a direct effect on the reverse current.

Photoconductive mode decreases considerably the reaction time to incident photons, so PDs biased in the reverse direction are used as high-speed PDs.

On the other hand, the temperature dependency of the PD current, though, is a downside of the reverse bias PD. In this conductive mode, current is measured through the circuit, which indicates the level of device's illumination, is direct and linear proportional to the input optical power. The width of the depletion junction, in

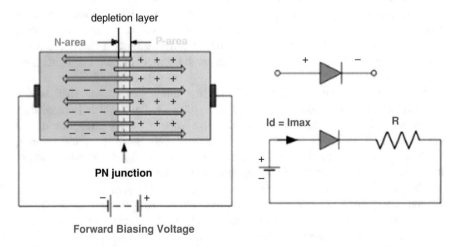

Fig. 2.11 Internal structure and symbol of a p–n junction diode

this case, increases, having a direct response in a high responsivity of PD, with a decrease in junction capacitance, producing a very linear response. However, these conditions are likely to produce a high dark current that can be limited based on the PD's material.

This dark current is produced by a leakage current flowing when a bias voltage is applied on PD. In photoconductive mode, the dark current is high and has significant fluctuation with the temperature' variation. For every $10°C$ temperature increase, the dark current doubles and the shunt resistance doubles with $6°C$. A higher bias increases the dark current present and decreases the junction capacitance. Both the PD's size of the active area and its material influence the dark current. Silicon devices, for example, have low dark current compared with germanium devices being therefore the favorite choice for VLC.

A PIN PD is a highly linear, fast device, having high quantum efficiency. The overall performance of the oRx front-end device depends on the bandwidth, sensitivity, and active surface of the PD as well as the quality of the communication channel.

An optical filter is placed in the system in order to select the spectrum of light that is sent to the receiver. The overall performance of the optical receiver (oRx) depends on the bandwidth, sensitivity, and surface of the detector as well as the quality of the communication channel.

When incoming light is absorbed in the depleted region of the junction semiconductor, it generates photocurrent, thus converting light into an electrical signal which is pre-amplified and then sent to data recovery and signal processing.

The PD is, therefore, an optoelectronic transducer generating an electrical signal that is proportional to the square of the optical signal striking on its active area. Hence, the electrical signal generated by the PD is proportional to the optical power received.

In order to achieve the expected data rate, the PD must have high quantum efficiency (η); high responsivity ($\mathscr{R}$); high photosensitivity (S) within its operational range of wavelengths; low noise level; minimum response at wide range of temperature's fluctuations; and long operational lifetime.

The ratio between the average number of electron–hole (e–h) pairs generated by a PD and the average number of incident photons in a certain time is named the *quantum efficiency* (η). Assuming that the absorption coefficient of the p–n *junction* is $\alpha(\lambda)$ and the junction width is w, then the quantum efficiency η is:

$$\eta = \frac{\frac{I_p}{e}}{\frac{P}{hv}} = 1 - e^{\alpha(\lambda)w} \tag{2.25}$$

For a high value of quantum efficiency (η), junction (w) must be as wide as possible. This is the reason why the I-layer is introduced between p and n layers.

The responsivity ($\mathscr{R}$) is defined by the ratio between the generated photocurrent (I_{PD}) and the incident power (P_0) at a given wavelength:

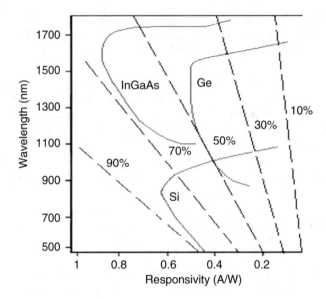

Fig. 2.12 Quantum efficiency (percent) of PDs according to their material [39]

$$R = \frac{I_{PD}}{P_0} = \frac{\lambda}{1.23985} \left[\frac{\mu m}{\mu m \cdot \frac{W}{A}} \right] \qquad (2.26)$$

Therefore, the spectral response is related to the amount of current and wavelength, as all wavelengths are at the same level of light.

The ratio of light energy in watts [W] incident on its surface to the resulting current in amperes [A] is called *photosensitivity*.

A small surface of a detector leads to low capacitance and high bandwidth. In case of a large surface of the PD, the power received increases by the same factor and is equal to an increase in sensitivity.

The spectral response as well as the quantum efficiency depends both on the material properties and PD's structure. The quantum efficiency (η) of the PD from the n and p layers can be ignored when the PD's surface-active area has zero reflectivity. In this case, the I-layer absorbs the entire light in the working voltage and therefore, the η is:

$$\eta = 1 - e^{-\alpha(\lambda)w} \qquad (2.27)$$

where:

$\alpha(\lambda)$—the absorption coefficient

w—thickness of the I-layer

The thickness of the *I-layer* (w) should be sufficiently large to improve η.

The long-wavelength limitation is influenced by the upper cut off wavelength.

Quantum efficiency (percent) of PDs according to their materials is shown in Fig. 2.12 where the dependency of the responsivity related to the wavelength can be seen. Si material is the best choice for 500–900 nm band while Ge and InGaAs are best suited for 1300–1600 nm bandwidth.

Quantum efficiency and responsivity vary with the wavelength. The quantum efficiency of the PD can be improved by reducing the reflectivity of the incident surface in order to force as many possible photons to move into the *pn* junction. As a result, when the width of the depletion region is increased, photons are completely absorbed in this region.

The current generated by PD without light on the active area is named the dark current. The Si-PIN PD has a minimum dark current, and on the other hand, the Ge-PIN PD has a maximum dark current. This is one of the main reasons why the Si-PIN PD is preferred in VLC systems.

Both the frequency and response time of a PIN PD are mostly determined by the transit time of the production of carriers in the depletion layer as well as the RC constant of the oRx front-end device including PD.

The transit time can be decreased by reducing the width of the depletion layer (w), thus improving the cutoff frequency. However, in this case, the quantum efficiency η will be reduced accordingly.

The circuit's RC time constant limits the cutoff frequency (f_c) as:

$$f_c = \frac{1}{2 \cdot \pi \cdot R_c \cdot C_d} \tag{2.28}$$

where:

R_c—sum of the series resistance and the load resistance of the PD.

C_d—sum of the junction capacitance (C_j) and the distributed capacitance:

$$C_j = \frac{\varepsilon \cdot A_j}{w} \tag{2.29}$$

where:

ε—dielectric constant

A_j—the junction area

w—the depletion layer width

The noise of a PIN PD mainly refers to the *shot noise* and *thermal noise*. Shot noise comes from both signal current and dark current. The thermal noise is produced by the load resistance and the input resistance of the amplifier.

Root mean square of the shot noise is:

$$RMS\, n_s = \sqrt{2 \cdot e \cdot I_p} \tag{2.30}$$

where:

I_p—peak current.

e—elementary electron charge $e = 1.602 \times 10^{-19}$ [C] (coulombs) [67].

Root mean square of dark current noise:

$$RMS\ n_d = \sqrt{2 \cdot e \cdot I_d} \tag{2.31}$$

where:

I_d—dark current.is:

$$RMS\ n_t = \sqrt{\frac{4 \cdot k \cdot T \cdot B}{R_r}} \tag{2.32}$$

where:

k—the Boltzmann's constant (1.38×10^{-23} [J/K]) [68].

T—equivalent noise temperature.

B—bandwidth.

R_r—parallel result of the load resistor and the amplifier input resistor.

In the VLC system, the shot noise is produced by the natural and artificial light present indoor/underground.

In the oRx front end, the thermal noise mainly comes from the electronic preamplifier i.e. the TIA. (The AD8015, is one example of TIA model that keeps a low AWGN power at the oRx with an important amplification of the optimum signal gained [39].)

The values of these two noise components (*shot noise* and *thermal noise*) are dominant at the oRx for the AWGN calculation.

At the oRx, the AWGN has a complex value with a double-sided electrical PSD. The electrical gain (σ_{AWGN}) of the oRx depends on wavelength as follows:

$$\sigma^2_{AWGN}(\lambda) = n_s + n_t \tag{2.33}$$

where:

n_s—shot noise

n_t—thermal noise

Shot noise is defined by equation:

$$n_s = 2e(P_o(\lambda) + P_{am}(\lambda)) \cdot R(\lambda) \cdot G_{TIA} \cdot G_{OC} \cdot B \tag{2.34}$$

where:

$\mathscr{R}$—PD's responsivity

G_{TIA}—TIA's gain.

G_{oc}—optical concentrators' gain

B—bandwidth

Thermal noise is defined by equation:

$$n_t = 4 \cdot k \cdot T \cdot B \tag{2.35}$$

In a VLC system, carriers with wavelength between 375 and 780 nm are useful signals, thus, all the other carriers in near IR, for example, can also be classified as noise.

Since the peak response of the PIN PD made by Si is between 625 and 900 nm (although the VLC stands around 420–760 nm) the communication signals are drowned by a quite high background noise induced by other types of light (natural or artificial light).

In case of a Si-PIN PD, the recombination of the photo-generated carriers in p layer and n layer reduces considerably its internal quantum efficiency (IQE). When light hits the n layer, high doping occurs and the PD carrier lifetime is significantly reduced and therefore its IQE for short wavelengths is low. On the other hand, in case of long wavelengths, the recombination of the photo-generated carriers in n layer doesn't matter much, due to its great penetration depth. Light with long wavelengths penetrates deep into silicon p layer; therefore, the recombination reduces the IQE.

Noise (forward-biased or shunt resistance noise) related to the shunt resistance (SR) is defined as the ratio voltage to the amount of current generated.

Noise equivalent power is the amount of light (of a given wavelength) that is equivalent to the noise level.

The SNR of a PD is defined as:

$$\frac{S}{N} = \frac{I_p^2}{2e \cdot (I_p + I_d)B} + \frac{4k \cdot T \cdot B}{R_e} \tag{2.36}$$

where:

B—bandwidth.

T—temperature equivalent noise.

R_e—equivalent resistance.

There are mainly two types of PDs used in VLC systems:

- Positive intrinsic negative (PIN).
- Avalanche photodiode (APD).

A PIN diode has a wide area between the *p-type* and *n-type* area (typically heavily doped), consisting of an undoped intrinsic semiconductor. The intrinsic area provides a greater separation between the p and n areas, allowing higher reverse voltages to be tolerated. The wide intrinsic area makes a PIN different from an ordinary p–n diode, being therefore suitable for photodetectors (Table 2.2).

APDs are high-sensitivity, high-speed semiconductor light sensors being widely used to convert optical data into electrical form. The main advantage of an APD is that it has a greater level of sensitivity compared to PIN since the avalanche action multiplies the gain of the diode.

Table 2.2 The working characteristic parameters of different material PINs

Parameters Symbol Unit	Si	Ge	InGaAs
Wavelength range k [nm]	400–1000	800–1650	1100–1700
Responsivity R [A/W]	0.4–0.6	0.4–0.5	0.75–0.95
Dark current I_d [nA]	1–10	50–500	0.5–2.0
Rise time τ [ns]	0.5–1.0	0.1–0.5	0.05–0.5
Bandwidth B [GHz]	0.3–0.7	0.5–3.0	1.0–2.0
Bias VB [V]	5	5–10	5

On the other hand, an APD needs a higher operating voltage and the output is nonlinear, therefore, APDs also have a higher level of noise than a PIN PD. Also, APDs, not like PIN PDs, need a high reverse bias state to work. That permits avalanche multiplication of the electrons and holes formed by the initial electron-hole pairs.

APDs have an internal area where electron multiplication occurs by use of an external reverse voltage. The gain resulted in the output signal means that, at high speed, low light levels can be measured. APDs are not as widely used as PIN PDs.

PINs are widely used where the resulting gain is important and have the advantage that the necessary working voltage spreads are between 5 and 15 V.

A PIN PD *responsivity* refers to its ability to convert optical power to electrical current according to the material it is made of being different for each wavelength. PIN's responsivity ($\mathscr{R}_{PIN}$) is defined as follows:

$$R_{PIN} = \frac{\eta \cdot e}{h \cdot f} \quad \left[\frac{A}{W}\right] \tag{2.37}$$

where:

η—quantum efficiency.

e—electron charge (1.6×10^{-19} C).

h—Planck's constant (6.62×10^{-34} J or $4.135 \ 10^{-16}$ eV s/rad).

f—frequency corresponding to the photon wavelength.

InGaAs PIN diodes have good response to wavelengths corresponding to the low attenuation window of optical fiber close to 1500 nm. The atmosphere also has low attenuation into regions close to this wavelength.

APDs are ideal for detecting extremely low light levels. This effect is shown in the gain G_{APD}:

$$G_{APD} = \frac{I_G}{I_P} \tag{2.38}$$

where:

I_G—value of the output current, amplified due to avalanche effect.

I_p—current without amplification.

APD has a higher level of the output current than PIN PD for the same value of optical input power, but the noise increases correspondingly by the same factor and, in addition, has a slower response than the PIN PD.

The spectral responsivity ($\mathscr{R}_{APD}$) that appears multiplied by G_{APD} in an APD is:

$$R_{APD}(\lambda) = \frac{I_G}{P_0} = \eta \frac{q \cdot \lambda}{h \cdot c} \cdot G_{APD} \tag{2.39}$$

Table 2.3 shows some of the materials and their physical properties used to manufacture PDs [69].

APDs are PDs with internal gain produced by the application of a reverse voltage. APDs have a higher signal-to-noise ratio (SNR) than PIN PDs, a faster time response, lower dark current, and higher sensitivity. The spectral response range is typically within 200–1150 nm.

Therefore, to create the optimum receiver design is a real challenge. Light emitted by LED is concentrated using optical elements both on the oTx and oRx, then is filtered and finally received on the PD's surface.

To achieve high data rate, the bandwidth-limiting effect generated by the yellow layer of phosphor has to be avoided.

In order to avoid it, there are some techniques used today in a VLC system:

- use of complex modulation schemes involving multiple bits carried by each symbol transmitted. This method involves merging multilevel modulation techniques like QAM with optical OFDM or DMT modulation. When used with blue filtering, the transmission rate can be extended to hundreds of Mbps [70],
- blue filtering at the PD to clean out the yellow components with slow response [71],
- preequalization at the oTx and post equalization at the oRx [72].

The choice of the best-suited PD for a specific VLC setup refers to its active area, capacitance, the spectral response as well as any transit-time limited bandwidth effects.

Table 2.3 Characteristics of PDs used in OWC systems

Material and structure	Wavelength (nm)	Responsivity (A/W)	Gain	Dark current	Rise time (ns)	Speed	Cost
Pin Silicon (Si)	300–1100 (visible to NIR)	0.5	1	Low	0.1–5	High speed	Low
Pin Indium gallium arsenide (InGaAs)	1000–1700 (NIR to MIR)	0.9	1	Low	0.01–5	High speed	Moderate
APD germanium (Ge)	800–1300 (NIR)	0.6	10	High	0.3–1	Low speed	Low
APD Indium gallium arsenide (InGaAs)	1000–1700 (NIR to MIR)	0.75	10	Low	0.3	High speed	Moderate

NIR near InfraRed, *MIR* middle InfraRed

PINs, APDs as well as single-photon avalanche detectors (SPADs) have the main role in VLC setups, using modulation schemes as pulse-position modulation (PPM) or PAM [73] as well as OFDM [74].

Although SPADs, as photon-counting detectors, have significant higher sensitivity than APDs for the same active area, there are difficulties in using them in VLC setups because of the fast dead time, high dark noise, and therefore a limited gain [75].

The most frequently used PD in VLC setups is the Si PIN PD due to its low voltage operation, linear response characteristics, high tolerance to a wide range of fluctuations in temperature, and low cost.

The Optical Filter

The optical filter used in a VLC system has to eliminate the light from any other natural or artificial sources; cutout the IR wavelength; remove the slow light from the yellow phosphor.

The yellow phosphor layer has been applied to the blue light in order to obtain white light in on-the-shelf commercial LEDs.

The optimum spectral response (SR) of the oRx depends both on the SR of the PD and the filter in front of it. As mentioned before, the SR of the PD is described by the $\mathscr{R}$, the ratio between the optical power and electrical current according to optical spectral range. Most of the on-the-shelf PDs $\mathscr{R}$ used for VLC setups today are between 0.6 and 0.8 A/W. The PDs SR raise from 320 to 1100 nm, therefore, optical filters have the role to separate the individual optical channels.

Optical filters used for VLC setups have a narrow optical bandwidth of 3 dB (10 ± 2 nm) in spectrum of UV/visible band-pass filters from 340 to 694.3 nm, with 1, 3, 10, or 40 nm band-pass regions and a gain factor of 0.8–0.9 (*Thorlabs* band-pass filters for visible spectrum e.g., FKB-VIS-10) [39].

The structure of a band-pass filter consists of several layers (dielectric stacks) of material on the surface of the substrate (spacer layers) (see Fig. 2.13).

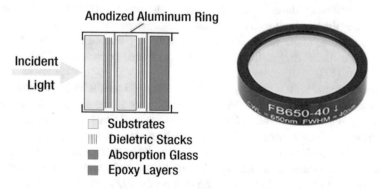

Fig. 2.13 Optical band-pass filter's internal structure and final product [76]

Each dielectric stack has a thickness of $\lambda/4$, where λ is the central wavelength of the band-pass filter (i.e., the wavelength with the best transmittance through the filter).

Each spacer layer (consisting of colored glass, epoxy, dyes, metallic, or dielectric), placed between the dielectric stacks, has a thickness of $(n\lambda)/2$, where n is an integer.

A Fabry–Perot cavity is formed by each spacer layer sandwiched between dielectric stacks. These interference conditions of a Fabry–Perot cavity resulted from its construction, allowing light to be sent efficiently at the central wavelength, and a small band of wavelengths, since all interference with negative effects stop the light outside the passband to be transmitted. Still, the band of blocked wavelengths on either side of the central wavelength is quite small.

Different materials with wide-blocking ranges are used between the spacer layers and the substrate in order to increase the blocking range of the filter.

While these materials successfully block out the transmission band of the incident radiation they also reduce the communication through the filter in the passband.

Filter orientation in the VLC setup is important to be done according to manufacture marks to avoid any thermal negative effects or possible damages. About 1 nm over the operating temperature range of the filter can result in shifts from the central wavelength of the band-pass filter [76].

The Optical Concentrator

The gain of a conventional optical concentrator (G_{oc}) is limited by its geometric properties that define the capability of the flux collecting of the optical system, called etendue. Etendue, in an optical system, is the property of light that characterizes how the light is "spread out" in both area and angle. Étendue is preserved as light travels through an optical system where it experiences reflections or refractions.

Etendue conservation means that gain of this optical concentrator can only be increased at the expense of decreasing the PD FoV. This limitation can be overcome when an optical concentrator is used. Light incident on the entrance aperture on the concentrator reaches the exit plane after being reflected on its side walls. The reliable concentrator for VLC setups has a hollow conical shape with reflecting material or a solid paraboloid, made from dielectric material. A paraboloid concentrator has a higher gain than the hollow one and is more commonly used in VLC applications [77].

A concentrator with a large gain in front of the PD results in an improved SNR since a small active area of the PD is preferably due to the small capacitance and therefore a higher bandwidth. Hence, an improved oRx has a concentrator with a large collection area, a wide FoV, and a PD with a small active area. The gain of the concentrator is limited by the condition imposed by conservation of etendue.

The optical concentrator usually receives light from a wide FoV (generally 2π steradians) and has the role to concentrate it on the PD's active area, therefore, its geometrical gain (G_{oc}), is:

$$G_{oc} = \frac{n^2}{sin^2\varphi} \qquad (2.40)$$

where:

n—the refractive index of the concentrator.

φ—half-angle FoV [78].

A steradian (sr) is defined as the solid angle subtended at the center of a unit sphere by a unit area on its surface. For a general sphere of radius r, any portion of its surface with area $A = r^2$ subtends one steradian at its center.

When a PD has a large active area, it collects a large amount of the emitted optical signal with the expense of a reduced electrical modulation bandwidth because of a high capacitance of PD. Instead of using a single PD, an array of PDs, collocated in a nonplanar mode (an angular diversity receiver, e.g., can be fabricated by assembling multiple small PDs on a semispherical shape base), eases this trade-off, increasing in this way the photosensitive area without reducing the modulation bandwidth [39].

2.2.2   Electrical Setup

oTx Driver

Hardware design and manufacturing of a suitable electrical transmitter module for a VLC setup is not an easy task since there is necessary to reach a proper balance between different optical and electrical characteristics considering that the electronic design has conflictual requirements.

Since LEDs emit noncoherent light and therefore, an unstable, nonlinear carrier, VLC setup employs the IM/DD method where the underground channel offers direct-data connection used as wireless communication link. The LED's nonlinearity can severely alter the output signal due to the large peak to average power ratio (PAPR).

One of the main requirements regarding the LEDs characteristics for a reliable optimal wireless communication is its fast response time. The LED, as the current-driven simple conducting device, has the brightness proportional to the forward current.

Taking into account that the main role of the LED in a fixture is illumination, not data communication, the oTx has to be designed to prevent LED's damage since the high current and therefore overheating would shorten its life. To ensure both a long time and normal operation of the LED and efficient VLC setup, some important requirements are necessary when designing the transmitter driver: its input DC voltage drop should not be lower than the LED' forward voltage drops; current must be controlled into the linear region; the driving circuit should adopt a DC current source or a unidirectional pulsed current source, rather than a voltage source [79].

Fig. 2.14 LED series resistor circuit with DC power supply

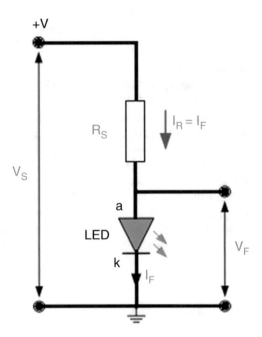

The LED, as the main actor of the oTx in a VLC setup, is a semiconductor device that converts—when operated in a forward-biased direction—the electrical energy into light. Because the LED's junction does not truly emit a high amount of light, its shell is covered inside by a transparent, hard plastic epoxy resin hemispherical shape that enforce the emitted photons to be reflected away and then focused upwards through the hemispherical top of the LED. In this way, the dome shell acts like a lens, focusing the amount of light.

The color of light emitted by the LED depends on the wavelength of the light being determined by the type of semiconductor compound used in the PN junction during its manufacturing process. White and blue light LEDs result by mixing two or more complementary colors at a precise ratio within the semiconductor composite and furthermore by inserting nitrogen atoms into the crystal structure during the doping procedure [80].

LEDs are devices dependent by current with forward voltage drop V_F, depending on the semiconductor structure (light color/wavelength) and the forward-biased LED current (I_F), as well. Most LEDs require a forward operating voltage from 1.2 to 3.6 V with I_F from 10 to 30 mA.

Although both the current and forward operating voltage fluctuate depending on their composition and dopant materials, for a regular red LED the step where conduction begins (and therefore light is produced) is about 1.2 V (3.6 V for a blue LED and 3.3 V for white LED).

The LED's light intensity is directly proportional to the forward current that flows through it. When connected in forward bias scenario across a power supply, LED must be current limited using a series resistor as seen in Fig. 2.14.

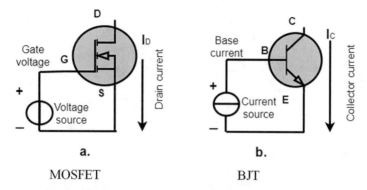

a.   b.

MOSFET   BJT

Fig. 2.15 MOSFET and BJT

The series resistor value (Rs) is calculated according to Ohm's Law based on the LED forward current (I_F) according to (2.45).

$$R_s = \frac{V_s - V_F}{I_F} \qquad (2.41)$$

In case of a 5 V DC power supply, when the red LED's current must be limited to 10 mA, for example, Rs has to be 380 Ω. To increase the LED's light intensity, its I_F should rise (e.g., to 30 mA) and therefore the Rs should be 126 Ω.

However, the brightness of a LED shouldn't be controlled by simply varying the current flowing through it because high values of current will result in more heat dissipation and thus a shorter LED's lifetime.

Besides LEDs, the most frequently used active devices for oTx driver are metal-oxide semiconductor field-effect transistors (MOSFETs) and bipolar junction transistors (BJTs), both of them having the same operating principle, being charge controlled devices, meaning that their output current is proportional to the charge established in the semiconductor by the control electrode.

The etymology of the word Transistor describes their mode of operation and is a combination of *Transfer* and *Varistor*. MOSFET is a voltage-driven device also called insulated-gate device. BJT is a current-driven device.

MOSFETs (Figure 2.15a) and BJTs (Figure 2.15b) are the key switching components in applications with high frequency and efficiency requirements, as VLC setups are [81].

In case of MOSFET, when a *voltage* is applied between the *gate* (G) and *source* (S) terminals, they produce a flow of *current* in the *drain* (D).

On the other hand, in the BJT case, a specific *current* must be applied between the *base* (B) and *emitter* (E) terminals to produce a *current* in the *collector* (C) (Fig. 2.16).

A MOSFET has low transfer resistance, being capable to operate at high current values and undertake low power dissipation. On the other hand, BJEs have the input

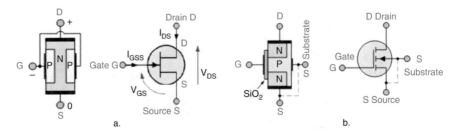

Fig. 2.16 (a) N-channel JFET and (b) N-channel MOSFET

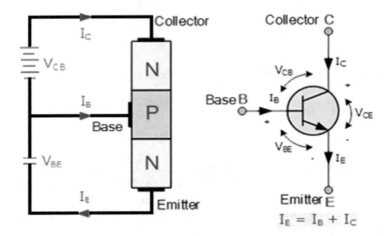

Fig. 2.17 NPN BJT

maximum voltage specification much lower and need to have high base current for proper operational phase and therefore lower input resistance than MOSFETs.

The DC–DC converter, used in some oTx driver setups, inputs an irregular DC voltage and outputs a constant or regulated voltage. The constant voltage source can be realized by using any of the conventional power converter topologies such as buck (e.g., EN6360QA or TMS320F2803x), buck-boost, and single-ended primary-inductor converter.

BJTs are devices regulating current handling the amount of current flowing through them from the **E** to **C** in proportion to the value of biasing voltage applied to the **B** terminal (Fig. 2.17).

As a small current flows into **B** terminal, controls a higher collector current forming the transistor action.

MOSFETs as well as BJTs have switching speed very close due to the necessary time for the charge carriers to cross the semiconductor region.

MOSFETs' most important characteristics are compared to BJTs' characteristics [80] in Table 2.4.

So far, two different categories of drivers for oTx have been developed:

Table 2.4 Comparison between MOSFET and BJT characteristics

MOSFET	BJT
Voltage-controlled device	Current-controlled device
Low voltage gain	High voltage gain
High current gain	Low current gain
Very high input impedance	Low input impedance
High output impedance	Low output impedance
Low noise generation	Medium noise generation
Fast-switching time	Medium switching time
Easily damaged by static	Robust
Some require an input to turn it "OFF"	Requires zero input to turn it "OFF"
Difficult to bias	Easy to bias
More expensive than bipolar	Cheap

Fig. 2.18 LED and T in series

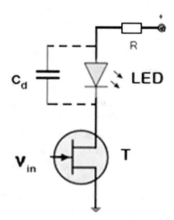

A. *Digital* oTx drivers (On/Off)—for digital modulation formats transmission.
B. *Analog* oTx drivers—for different more complex modulation formats being necessary continuous output level or multiple output levels.

Digital oTx Drivers (ON/OFF)

In case of a digital oTx driver that allows different modulations, the current control is coming from an input signal with a stable voltage and low current capabilities.

Fig. 2.19 LED and T in parallel

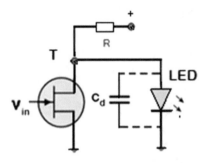

In the circuit shown in Fig. 2.18, when V_{in} rises, both the current in transistor (T) and LED rise. In case that the voltage across the T is considerably smaller than the LED forward voltage, the current is limited by R.

For a high rate switching, C_d must charge and discharge at a high rate.

When current rises in T, C_d starts charging, being though limited by R.

The moment when T switches off, C_d has a slow discharge due to a high resistance value.

A solution to the slow discharge of the C_d is to connect in parallel a second T.

A LED transistor driver circuit is shown in Fig. 2.19.

A different configuration is the one in Fig. 2.19 where LED is active at the moment when V_{in} is low. The LED's capacitance is discharged through $R_{SD(on)}$ while charged through R, therefore, there is not a proper balance between rise/fall times. $R_{SD(on)}$ is the total resistance in the path from source to drain, crossing the path of current flow.

Another useful LED oTx driver circuit is based on a complementary (*n-type* and *p-type*) metal-oxide semiconductor (CMOS) inverter (Fig. 2.19).

Considering an ideal situation, when the devices are perfect, without leakage current, in the stationary phase, the circuit does not consume power.

Both gates of Q_1 and Q_2 are at the same bias V_{in}, meaning that they are in a complementary state all the time. When V_{in} is high and equal to supply voltage (+), the *n-type* transistor is ON, while the *p-type* is OFF. Alternatively, when the input voltage is low (0 V), Q_2 is OFF, and Q_1 is ON. The Q_1 transistor feeds current to LED and charges C_D. The second transistor, Q_2 drains the charge from C_D (current sink) (Fig. 2.20).

When LED dimming is necessary, less current (let's say, e.g., below 5 mA), LED will dim its light output significantly or even will turn it OFF completely. A most proper way to control the LED's brightness is to use PWM when LED is turned ON and OFF repeatedly at varying frequencies (Fig. 2.21).

The same PWM is used when higher light outputs are necessary using an equally short duty cycle—ON–OFF ratio. In this way, the output light intensity and therefore the LED current increase significantly, keeping, on the other hand, the power dissipation in limits that are safe. Pulses of the LED's ON–OFF at frequency of 100 Hz or more are brighter for the eye but still seem continuous.

Fig. 2.20 CMOS inverter

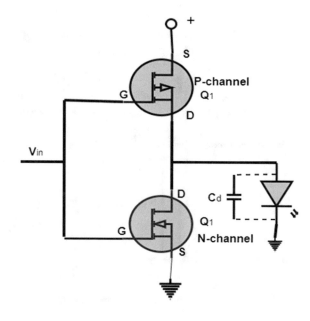

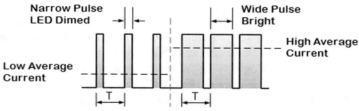

Fig. 2.21 Pulse width modulation current

PWM is used to encode a message into a pulsing signal. This modulation technique is used to encode information for wireless communication. The narrow pulse comes from the LED dimmed with a low average current where the brightness of LED results in a wide pulse due to a high average current.

A common amplifier circuit is shown in Fig. 2.22.

Analog oTx Drivers

More complex modulation techniques such as QAM or OFDM require an analog oTx driver. One important advantage of analog drivers is that they have high linearity.

There are also two different approaches here: signals can be represented by voltage or by current.

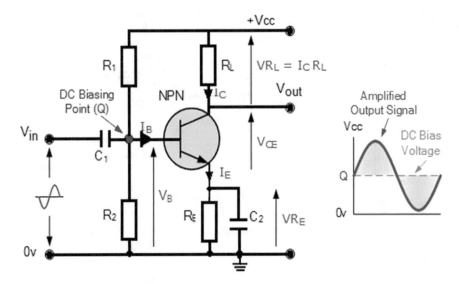

Fig. 2.22 A regular amplifier circuit (adapted from [80])

In case of a LED voltage-mode analog driver, there are linearity issues since the current/voltage characteristic is intrinsically nonlinear, and therefore, voltage/current conversion has nonlinear effects that drive to low performance for PAPR.

The current-mode analog drivers are more appropriate due to both a linear behavior of the LED's power/current ratio and the system's high speed [82].

The oTx driver circuit can operate at larger bandwidths than the LED's bandwidth limited at 3 dB. In order to overcome this inconvenient, the oTx analog driver circuit has a preequalization stage which is an efficient way to enhance the data rate and increase the bandwidth, as well.

Fujimoto et al. [83] achieved in 2013, a bit rate of 477 Mbit/s using OOK-NRZ modulation scheme, a low-cost commercial available PIN PD and an adapted LED driver with a simple preemphasis circuit with a BER less than 10^{-9}.

Another example of high speed, low complexity VLC circuit was designed in 2014 by Li et al. [84] using WLED analog preemphasis circuit at oTx and post-equalization at oRx.

The oTx driver circuit has three stages amplify circuit Q_1 and Q_2 BJT BFR520 and Q_3 BFR540, an amplifier type ZHL-6A, Bias-T type Aeroflex 8810, and LED type OSRAM LUW W5AM (Fig. 2.23).

The oRx circuit has a differential amplifier ADA4937–1, a TransImpedance Amplifier TIA type MAX3665, and a PD type HAMAMATSU S10784. It has been used as a simple Not Return to Zero—On–Off Keying (NRZ-OOK) modulation technique [85].

They extended communication from 3 to 233 MHz using a blue filter placed in front of the PD, achieving the highest performance in VLC systems by then. They

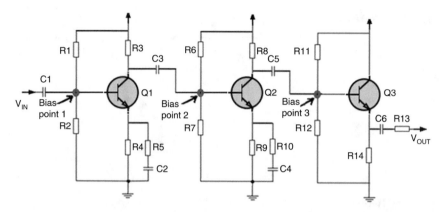

Fig. 2.23 Tx with preemphasis drive circuit with wideband NPN transistors (adapted from [85])

reached a data rate of 480 Mbit/s and even 550 Mbit/s at a distance of 60 cm and 160 cm, respectively.

The system designed had a BER of 2.6×10^{-9}, considerably lower than the FEC limit 3.8×10^{-3}.

Next year, in 2015, Huang et al. [86] demonstrated 750 Mbit/s using a bridged-T amplitude equalizer embedded in the oTx driver circuit with 64QAM-OFDM modulation scheme, based on an RGB LED and a high sensitivity APD PD with BER under pre forward error correction (FEC) limit of 3.8×10^{-3}.

Software equalization provides more accurate results. Zhou et al. [87] experimentally implemented a power exponential software preequalization and reached 2.08 Gbps based on a WLED, lens, and blue filter, at a distance of 1 m.

When LiFi is implemented in the lighting fixture, the most important issue that has to be taken into account is that the primary function of the setup has to be illumination and only the second one is data communication. Fast wireless data communication can be achieved with LEDs fast switching that can affect the illumination function. In order to accomplish a proper balance on both functions— illumination and wireless signal communication—the LiFi setup has to combine data wireless communication with LED biasing. The most used scenario to achieve this task, rely on a bias-T. A VLC bias-T (see Fig. 2.24) consists of passive LC components where input is applied at the capacitor, output is taken from the LC node and L is grounded.

For illumination purpose, the bias-T has to provide strong attenuation for signal, passing the low frequencies and for data communication, strong attenuation at low frequencies has to be achieved.

There have also been developed special circuits delivering a separate control of both functions, illumination, and data signal for wireless communication [88].

Fig. 2.24 VLC bias-T

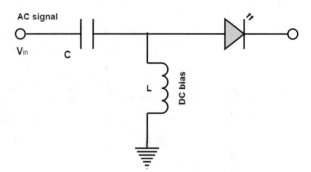

The oRx Driver

The PD converts all types of light (natural and artificial light) that hits its active surface area into a current that is proportional with the light's intensity.

At first thought, in order to detect an ideal optical signal sent by LED to PD through free space, the PD's surface-active area should be as large as possible. However, a large active area of the PD has a significant trade-off: large intrinsic capacitance and therefore a reduced electrical modulation bandwidth.

As an alternative solution, instead of one PD, at the oRx can be used an array of PDs in order to avoid this trade-off. A possible geometry of the PD matrix can be shaped to receive any pattern of light depending on different situations or special requirements [39]. Other different solutions consist of different frequency optimization procedures including capacitive [89] and inductive peaking [90], [91].

The PIN PD converts optical signals into current that is processed by TIA. The TIA front-end circuit amplifies the signal and converts the PD's current into voltage, with low penalties on gain, bandwidth, and noise [92].

The current produced by the PIN PD can be amplified with a reverse voltage. This solution is not appropriate since a reverse voltage would result in an increased reverse leakage current (also known as dark current) that creates additional unwanted noise. A good alternative is the one using the TIA which will amplify the signal and convert current in voltage, increasing therefore both gain and speed [93].

However, designing an appropriate TIA for a performant oRx in a VLC setup is not an easy process since there it is also difficult to establish a correct balance, a proper trade-off between gain and bandwidth. TIA used in VLC setups are classified as (1) open-loop TIA (with low input impedance amplifiers or high input TIA) and (2) feedback TIA [94].

1. Open-loop TIA

    Low input TIAs are appropriate for high bandwidth and low noise performance in VLC oRx setups with low sensitivity drawback. In contrast, high input TIA is highly sensitive with low-frequency performance.

Fig. 2.25 Feedback TIA

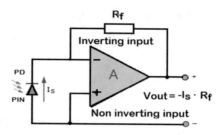

2. feedback TIA

On the other hand, feedback TIAs overcome the above-mentioned drawbacks of the open-loop TIAs having small input impedance with high bandwidth and high gains with low sensitivity [95].

The feedback resistor (R_f) sets the operating voltage level at the inverting input. The output voltage V_{out} is proportional to input current generated by PD (Fig. 2.25).

Achieving a proper trade-off between gain and bandwidth when a TIA is designed is not the only challenge, since noise has to be taken into consideration as well. Although there are many sources of noise in electronic circuits, (1) shot noise, (2) thermal noise, and (3) flicker noise are the most important, affecting the performance of the TIA in an oRx circuit of a VLC setup. Flicker noise (or 1/f noise), dominates at low frequencies and thermal noise dominates at high frequencies. Shot noise and thermal noise are white noise sources with small spectral density.

Data received by the PD from LED can be seriously compromised by noise. All additional light sources, besides the LED's light coming from the oTx, produce electrical current at the oRx. When the signal is weak, the SNR value is low, as well as BER, and therefore, error detection, equalization, or correction codes are methods applied to overpass the negative results of a weak optical signal [96].

In order to remove all additional unwanted signals that become noises, a filter has to be considered in the oRx hardware.

In order to reduce unwanted noises, both the shot noise and thermal noise, a low-pass Bessel filter is proposed by K. Sindhubala et al. [97].

TIA has to operate in low noise conditions and wide bandwidth of frequency. Adiono et al. present a detailed description of noises and make a deep analysis of frequency band in TIA's design in order to cope with VLC's high-performance requirements [98].

Low frequencies are undesirable noise resulting from optical excess signal, therefore, a high-pass filter (with a bypass capacitor) at the input of the front end is possible to be used. However, this method is not suitable for integration because a large area is necessary to implement a proper capacitor.

To remove the low-frequency noise, Chang et al. propose a fourth-order high-pass filter with Sallen–Key method consisting of two high-pass filters in cascade and a Schmitt trigger. They designed this new VLC oRx architecture based on OOK and experiments showed that the interference effects produced by the low-frequency noise of AWGN are considerably reduced achieving 12 dB gain [99].

Fig. 2.26 TIA

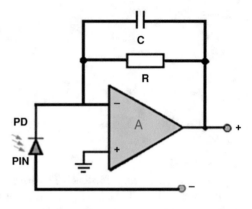

Fig. 2.27 Passive analog
high pass filter

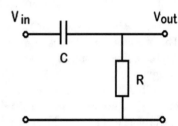

Karimi-Bidhendi et al. proposed a new TIA designed for a simplex wireless data VLC, consisting of three embedded stages:

(a) a shunt series feedback that has a transformer-based positive feedback module,
(b) an RC—degenerated common–emitter module,
(c) an inductively degenerated emitter follower module.

TIA developed achieves a measured gain of 41 dB and an input current noise spectral density of 39.8 pA/√Hz over a 50 GHz bandwidth [100].

A different, simple approach for a suitable oRx in a simplex VLC setup has been proposed by Böcker et al. consisting in a TIA (Fig. 2.26), high-pass filter (Fig. 2.27), an automatic gain control circuit (used to keep signal strength to different oTx to oRx distances) followed by an analog to digital converter from front to end in this order [93].

2.2.3   Channel Models for IR and VLC

Due to the recent years of intensive research to find an alternative solution for the "spectrum crunch," many new VLC applications for indoor scenarios have been developed.

Regarding the channel model for optical wireless communication, some important research papers and books have been written [88], [101–105], several models in

MatLab, codes on GitHub [106], and different other advanced dedicated software have been so far developed. Since light in the visible spectrum has several particularities, the various scenarios of VLC are not deeply investigated and determined, yet.

Not even the IEEE 802.17.5 standard for VLC [107] does not specify the channel's model to be used for evaluation yet, thus a universal channel model that can be applied to a wide range of indoor or underground scenarios has not been published in the present literature.

Although a growing interest in the wireless VLC technology has been recently noticed, the channel models for visible light are still not entirely characterized, yet. This is a challenging issue since the visible light channel model is one of the most important studies to be done when a reliable, efficient, and robust VLC system is intended to be designed [108].

Between IR and VLC wireless communication, there are significant differences, therefore, many specific characteristics of VLC have to be taken into consideration, when the channel model has to be accurately described.

The broadband nature of visible light (between 380 and 780 nm) makes the channel models for well-known narrow-band IR wireless methods unsuitable for a direct application of them in VLC [109].

So far, the studies presented for indoor OWC (both VLC and IR) channel models, can be classified as they can be seen in Fig. 2.28, in deterministic and stochastic models.

The *deterministic optical channel models* developed so far are typically based on the detailed description of specific propagation environments such as the optical channel topology with exhaustive explanation of the scenario (the objects indoor and their characteristics), as well as the position and orientation of both oTx and oRx. The CIR $h(t)$ of the OWC system is obtained using rigorous simulations that incorporate most of the details of the indoor propagation environment. All these models are site-specific, being physically meaningful and potentially accurate.

The deterministic approaches investigated till now, such as the recursive [46], iterative [110], ceiling bounce [111], DUSTIN [112], and ray tracing GBDM [88], [102], [113] are examples of geometry based deterministic models [114–116].

The *recursive* approach is used to solve a given problem by breaking it up into smaller pieces, solve it, and then combine the results. The first investigation evaluated the CIR $h(t)$ scaled by time domain, taking into consideration multiple bounces (more than two reflections) in wireless IR channels.

This approach, investigated by Barry et al. [41] followed the single reflection cavity model previously proposed by Gfeller and Bapst [40]. In this model, the radiation intensity pattern $R(\varphi)$ for a particular oTx can be modeled using a generalized Lambertian radiation pattern as shown in Eq. 2.6. Therefore, the received optical signal is proportional to the active area of the PD multiplied by $\cos(\omega_i)$, where ω_i is the incident angle of the oRx. Only rays that are incident within the FoV of the oRx are going to be captured, hence being the carriers of the useful optical signal. In this case, the CIR is the superposition of the LoS topology where an infinite sum of multiple-bounce components is added. The LoS response is approximately a scaled

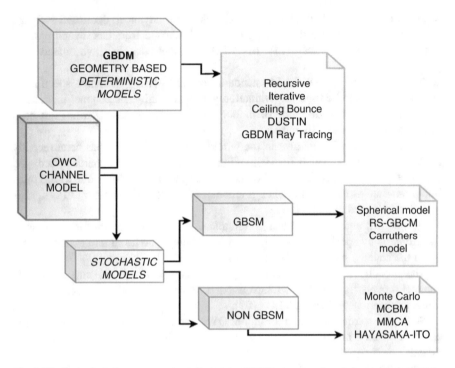

Fig. 2.28 Optical wireless communication models. *GBDM* geometry-based deterministic models, *GBSM* geometry-based stochastic model, *NON GBSM* non geometry-based stochastic model, *RS GBSM* regular-shaped GBSM, *MCBM* Modified Ceiling-Bounce Model (CBM), *MMCA* Modified Monte Carlo Algorithm

and delayed Dirac delta function ($\delta(.)$) expressed as in Eq. 2.14. The implementation of this algorithm is done by breaking all the considered reflecting surfaces of the objects inside the environment indoor, into numerous small Lambertian reflecting elements, called cells. These cells, individually considered with their own area, play the role of both an elemental receiver and an elemental emitter. The recursive approach accuracy increases with the number of the reflections considered but the computing time increases exponentially, as well. The downside here leans on the constant overestimation of the optical path loss and the bandwidth when too few reflections are taken into account [113].

In the paper [110], the CIRs are generated based on geometrical modeling of indoor environments together with an *iterative* technique for calculating multiple reflections. In this work, the authors follow the basic methodology described in [45] and additionally split the environment into many boxes. They showed in this way, that when three bounces of the reflected rays are considered, the iterative method is about 92 times faster compared with recursive method [110].

The first *ceiling bounce* model proposed was dedicated to the diffuse indoor IR communication, being distinguished by simplicity and the close accuracy between simulation and the data measured. The model is more realistic with multiple reflections and considers both the vertical separation between oTx and oRx and the distance between the oTx and the ceiling when the oTx, fixed on the ceiling, points down to the oRx.

DUSTIN algorithm, developed for IR, is based on a discretization of the reflecting surfaces into cells and is similar to the iterative and recursive models [112]. Unlike the previous approach described here, this algorithm developed by authors aims to reduce de computational complexity, by splitting the reflections into time steps rather than the number of reflections, allowing to perform simulations with any number of reflections. DUSTIN algorithm, although not tested on VLC yet, is considered faster compared with traditional iterative and recursive models.

The geometry-based deterministic models based on *ray-tracing* algorithm mainly rely on the theory of geometrical optics and the uniform theory of diffraction. Since it considers individually all rays for each receiving point and computes them, is the uppermost time-consuming prediction method. With the support of a ray-tracing dedicated application, the entire environment can be created considering different geometries of the furniture with particular materials (wood, glass, plastic, or metal) and their own reflection characteristics as well as the specifications of the oTx and oRx. These features of CAD integration are all included into the Zemax® OpticStudio software, for example. For all the objects incorporated, the reflection coefficients can be defined as a function of the wavelength according to each material on the objects' surface. Zemax® is a nonsequential ray-tracing tool that uses also the Monte Carlo algorithm, and allows to generate an output file, which includes useful data about rays such as the detected optical power as well as the path lengths for each ray. Due to its interoperability to MATLAB®, the output file can be imported to this software, where, using the information gained, the CIR value can be obtained, according to Eq. 2.42.

$$H(t) = \sum_{i=1}^{N_r} P_i \delta(t - \tau_i) \tag{2.42}$$

where:

P_i = the power of the ith ray.
τ_i = the propagation time of the ith ray.
$\delta(t)$ = the Dirac delta function.
N_r = the number of rays received at PD.

Studies indicated that, when furniture is considered in an indoor VLC topology, both the root mean square (RMS) delay spread (DS) and the DC channel gain H (0) decrease. Moreover, for the same site setup, in the case of the VLC, RMS DS and the DC channel gain values are poorer than with IR transmission.

The geometry-based deterministic model based on ray-tracing algorithm cannot be taken into consideration to be generalized to a wider class of scenarios for the optical channel model, since the propagation environment where the VLC setup is installed is potentially different for each indoor particular situation. On the other hand, the advanced tools offered by complex applications available on market, allow a careful design of a VLC setup for an accurate, and performant optical transmission in any particular case considered.

The *stochastic optical channel models* are based on the random behavior of the light waves propagation, resulting in the CIR applied into specific oTx and oRx, being predefined in a random approach according to certain probability distribution. These methods offer high flexibility, being non-site-specific and less complex computational methods, however, with relatively low accuracy.

The stochastic methods are split into geometry-based (spherical, regular-shaped, and Carruthers model) and non-geometry-based methods (Monte Carlo algorithm, modified ceiling-bounce, modified Monte Carlo algorithm, and Hayasaka-Ito model) as in Fig. 2.28 [104].

The *spherical* model is inspired by the traditional approach of the integrating sphere photometry. The most important outcome of using this method is that a highly reflective geometry drives to a high DS, thus a low channel bandwidth. On the other hand, a low reflectivity of the geometry leads to a low DS with a high channel bandwidth. The spherical model has been demonstrated to be reliable for IR communication in case of high order reflections, confirmed as well, by the ray-tracing simulations that take the diffuse reflections into account [114]. However, results obtained after simulation for a VLC transmission, showed that the CIR values of the diffused part were overestimated when this spherical model is considered since these diffused portions do not have an important influence on 3 dB bandwidth [115].

The *regular-shaped* channel model applied on optical channel has been inspired from the RF channel model where the effective scatterers are positioned on regular 2D shapes (ring, ellipse) or 3D shapes (sphere or ellipsoid). However, real scatterers can impose different delay due to various angles. For a VLC model, following two-rings and an ellipse model investigation [104], results showed that the values are comparable with RF studies for up to three reflections.

The *Monte Carlo algorithm* [116] follows three steps: first, the rays are generated, then, the walls' feedbacks are processed and last, the PD response is considered, allowing to determine the CIR not only for the Lambertian source, but for the specular reflections, as well. The computation complexity of this method is lower than in case of the DUSTIN model, however, in case of a regular-sized room, since not all the rays sent will hit the active area of the PD, the number of rays emitted by oTx must be high.

The *modified ceiling-bounce algorithm* applied also for IR communication includes additionally the contribution of the walls to the total value of the CIR.

The *modified Monte Carlo algorithm*, on the other hand, although first developed for IR transmission, has been applied based on the Lambert–Phong pattern for both single and multiple sources, and the simulations showed that the calculations become linear to the number of reflections and the computational complexity decreases.

A different taxonomy regarding the indoor optical wireless communication model techniques, exclusively for the VLC has been more recently defined by Ramirez-Aguilera et al. [108]. Here, VLC channel models are classified as analytical, iterative, and statistical.

The analytical methods investigated that describe the visible light channel model, are simple, and, although they offer an accurate value of the CIR, most of the scenarios investigated are ideal environments with unrealistic scenarios.

Unlike analytical models, the recursive or iterative channel models describe better the channel behavior, in different conditions where is realistic considered that the optical signal captured by the active area of the PD experiences time dispersion due to reflections from ceiling, floor, walls, and other objects, the most reflections being naturally diffuse.

The iterative methods refer to the propagation environment considered in small cells or a combination of the Monte Carlo simulations with the ray-tracing technique.

All the statistical techniques, that aim to estimate the optical CIR, are built on the different scenarios (without reflections or with one/two bounces of the light rays from walls or obstacles indoor) and various VLC topologies (LoS, NLoS, etc.). These statistical models, although they raise the computational complexity and require advanced and expensive applications for simulation, or, an extensive amount of data collected during experimental tests are the closest to actual situations and therefore the most reliable ones.

Besides a very clear and accurate description of the challenges that have to be overcome to define an accurate description of a general channel model for VLC, the authors in [108] propose a generalized simulation model of CIR for indoor VLC channels with the property of capturing reflection phenomena with changes on wavelength such as fluorescence, phosphorescence, or iridescence. They also propose a multiwavelength matrix-modified Monte Carlo model that allows to obtain data about the mechanism of surface reflections that can better capture the visible light channel characteristics.

According to Dimitrov and Haas, the visible light wireless channel is linear, memoryless, time invariant (exception being situations when light beam obstruction and shadowing occurs), with an impulse response of a finite duration [39, 117].

RMS DS and optical path loss (PL) are two characteristics that define the optical wireless channel. DS measures the multipath density of a channel of communication. It is the difference between the time of arrival (ToA) of the earliest significant component (usually LoS component) and ToA of the latest multipath components. The most used metrics for spread delay is RMS DS. The power delay profile (PDP) describes the intensity of a signal received (through a multipath channel) as a function of time. The abscissa is usually in units of time and the ordinate is in decibels (dB) [118].

For the modeling of the channel, the PDP of a channel is obtained taking into consideration the spatial average of the channel's baseband impulse response (C_bIR) $|h_b(t, \tau)|^2$ in a specific room.

RMS value of a signal ($x(t)$) is calculated as the square root of average of squared value of the optical signal, mathematically represented as:

$$E_{RMS} = \sqrt{\frac{1}{T} \int_0^T x(t)^2 dt} \qquad (2.43)$$

For a signal represented as N discrete sampled values—$[x_0, x_1, \cdots, x_{N-1}]$ the RMS value is given as:

$$E_{RMS} = \sqrt{\frac{x_0^2 + x_1^2 + \cdots x_{N-1}^2}{N}} \qquad (2.44)$$

When the signal is represented in frequency domain as $X(f)$, then, the RMS value can be calculated as:

$$E_{RMS} = \sqrt{\sum \left|\frac{X(f)}{N}\right|^2} \qquad (2.45)$$

The temporal dispersion of the optical CIR $h(t)$ can be defined by the channel RMS DS which is calculated based on the impulse response [47] as follows:

$$D = \sqrt{\frac{\int_{-\infty}^{+\infty}(t-\mu)^2 h^2(t)dt}{\int_{-\infty}^{+\infty} h^2(t)dt}} \qquad [ns] \qquad (2.46)$$

where:
μ—mean delay spread (DS) [47] is given by:

$$\mu = \frac{\int_{-\infty}^{+\infty} t \cdot h^2(t)dt}{\int_{-\infty}^{+\infty} h^2(t)dt} \qquad (2.47)$$

The value of channel RMS DS is used to compare the CIR width and its effect on the transmission system for different configurations, therefore, both $h(t)$ and RMS DS are fixed for a certain VLC setup configuration. Measurements showed that the RMS DS's value has significant importance in the power penalty due to inter-symbol interference (ISI) [47].

In case that the symbol duration is long enough relative to the DS, an equivalent inter-symbol interference (ISI)—free channel is expected. The larger the coherence bandwidth (CB), the shorter DS and therefore, CB is related to the inverse of the DS. Equivalent, the shorter DS, the larger CB. DS has a substantial impact on the ISI, too [119]. The multipath dispersion can cause ISI at oRx. ISI can be measured by the RMS DS. RMD DS is a measure of the temporal dispersion of the signal at oRx due to multipath propagation that causes ISI [120].

An important characteristic of a wireless optical multipath channel is that the channel stretches the transmitted signal in time, phenomena being known as *temporal dispersion.* All delays measured longer than the delay corresponding to the arrival of the first transmitted signal at the receiver is called excess delay (ED) (τ_i). Temporal dispersion (σ_τ) can be quantified by the channel RMS DS [121].

Optical path loss is defined as the ratio of the transmitted to received power, expressed in decibels (dB). The ratio of RMS DS and symbol time duration quantifies the strength of ISI. This ratio defines the complexity of the equalizer necessary at the receiver. Normally, when the symbol time period is greater than 10 times the RMS DS, there is not necessarily an ISI equalizer at the receiver [122].

For indoor VLC setups (empty room of about $5 \times 5 \times 3$ m), frequent values of RMS DS reported are between 1.3 and 12 ns for LoS links, between 7 and 13 ns for NLoS links, and the entire DS up to 100 ns for a channel optical path losses up to 80 dB [39].

For example, unlike an IR emitter, where the signal is considered monochromatic, the signal generated by VLC emitters can also be polychromatic. Moreover, IR communications assume that the reflectance of materials is typically modeled as a constant, but the wideband nature of VLC yields a wavelength dependency in the reflectance of materials.

The main reason why the VLC channel model is not determined yet is related to the high level of complexity that light's properties experience when traveling in different, complex environments.

Light has many intrinsic properties and when it travels through open spaces indoors, different phenomena can be observed such as refraction, reflection, dispersion, absorption, scattering, interference, diffraction, or polarization. Some of the scientific works that treat the VLC channel model subject focus mainly on the reflection phenomenon and ignore all the other relevant phenomena. Even when considering reflection, it is assumed that all the surfaces of the objects indoor have the same diffused reflection. This is far from an accurate description of the real behavior of light and the resulting evaluation of communication when light travels from an oTx to an oRx in a specific VLC setup.

Light, being an electromagnetic radiation, has the properties of waves. This intrinsic wavelength dependence in several phenomena is an important property.

When light "hits" an object' surface, its reflectance properties depend on many factors, such as the material of the object and its surface, the chemical composition and physical state of the object's surface, the texture (its roughness or smoothness) of the surface, the geometry of the object (important to calculate the incidence angle of light's beam), the color of the object's surface, and the structure of object's surface.

There are many studies defining different features of the objects' surfaces or materials by analyzing their spectral reflectance patterns specifically expressed in a form of curve as a function of light's wavelengths.

Since the refractive index of all kinds of materials depends on the wavelength of light, different wavelengths are interfered, to different extents, by the atoms into the material composition. Regarding the refractive index, as a broad conclusion due to many observations, has been established that the refractive index linearly varies with the light's wavelength. Because the refractive indices are different for each wavelength of light, the effect of light dispersion occurs. Since dispersion and absorption are closely related, a dispersive surface, with wavelength-dependent refractive index, must also be absorptive and, therefore, the absorption coefficient must be wavelength dependent.

Even though the optical signal may experience specular reflections when it hits smooth objects as a mirror, for example. For this reason, most of the reflections underground have to be considered as diffused reflections when the VLC channel model is described.

Hence, VLC channel models for the light propagation can be improved by, on one hand, incorporating all the physical phenomena that the light beam experiences when traveling underground from emitter to receiver and, on the other hand, the pattern of the power distribution has to be considered, in the entire spectrum of the visible light.

The light beam that propagates in polluted environments, as underground mine environments are, upon interaction with the tiny solid particles in suspension suffer from low to high attenuation because of both absorption and scattering phenomena.

Absorption refers to the energy transfer from the wave to the tiny solid particles of the underground polluted environment. Following this interaction, the energy transferred to the particles results in vibrations or rotations. The wavelengths of light in this case depend on the energy-level structures and therefore on the type of particles (molecules and atoms) present in the polluted environment. The spectrum of the light after passing through a polluted environment has certain wavelengths removed because they have been absorbed. It is possible that absorption would lead to a permanent loss of optical power as long as light propagates in high polluted environments. This occurs due to the photons' interaction with suspended molecules and particles found in the light's way from the oTx to the oRx. Absorption depends on the variation of the environment refraction index (n) and wavelength of the light (λ). Selective absorption has also to be considered underground due to black and grey dominant colors.

Scattering, defined as the deflection of light from its original path, is the light's property most evaluated into a polluted industrial environment. On the microscopic level, scattering relates to the interaction between a light photon and a molecule or an atom. Moreover, particles of different types of material with different shapes, concentrations, and humidity effectively determine the scattering properties of the environment.

Since scattering is the redirection of light caused by its interaction with matter, radiation may have longer or the same wavelength as the incident radiation.

If particles in air are much smaller than the wavelength (λ) of light, they absorb the incident light and quickly reemit the light in different directions.

In case that the reemitted light has the same wavelength (λ) as the incident light, the process is termed *Rayleigh* scattering (Figure 2.29a).

When the reemitted light has a longer wavelength (λ), the molecules are left in an excited state and the process is termed *Raman* scattering. In Raman scattering, secondary photons of longer wavelength are emitted when the molecule returns to the ground state (Figure 2.29b).

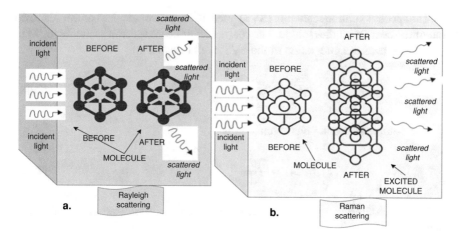

Fig. 2.29 Rayleigh scattering (**a**) and Raman scattering (**b**)

According to literature, blue wavelength is scattered more than the yellow one. As opposed to sound, the speed of light is slower in water than in air. Molecules of air O_2 and N_2 are Rayleigh scatterers for visible light and are more effective at scattering shorter wavelengths (such as blue and violet).

In fact, even in small concentrations, these particles make the scattering highly peaked in the forward direction, which is one of the major characteristics of the visible light propagation in a polluted environment.

The performance of an underground VLC (UVLC) system is highly affected by channel fading as a result of air movement with tiny solid particles in suspension.

This is similar to the atmospheric turbulence in FSO communication when fog and rain are also considered. Spots of polluted, movement air, with suspended particles with different shapes and sizes continuously change the propagation direction of photons due to the variation of refraction index, n [123].

Light properties into underground mining environment can be described as intrinsic and apparent.

Light's intrinsic optical properties (IOPs) depend exclusively by the optical medium when the apparent optical properties (AOPs) are dependent both on the optical medium and the environment studied. The environment studied refers to the surrounding within its space particularities: additional natural and/or artificial light sources, geometrics, type of materials, and color of the objects' surfaces within the space where the VLC setup is considered. The IOPs are conservative properties and hence the magnitude of the absorption coefficient linearly varies with the concentration of the absorbing material. Theoretically, the absorption coefficient can be expressed as the sum of the absorption coefficients of each component in the medium [124].

The two intrinsic optical properties in underground mines (IOPUMs) that model light absorption and scattering are both the function of spectral volume scattering $\beta(\theta,\lambda)$ and the coefficient of spectral beam absorption as (λ) in $[m^{-1}]$.

The spectral volume scattering (SVS) refers to the part of incident power scattered out of the beam with the θ angle. The coefficient of beam spectral scattering $b_s(\lambda)$ in $[m^{-1}]$ results as the integration of the SVS in all directions:

$$b_s(\lambda) = 2\pi \int_0^\pi \beta(\theta, \lambda) \sin\theta \, d\theta \qquad (2.48)$$

The volume scattering phase function is:

$$\widetilde{\beta}(\theta, \lambda) = \frac{\beta(\theta, \lambda)}{b_s(\lambda)} \qquad (2.49)$$

The spectral beam attenuation coefficient $c_a(\lambda)$ considered also as optical power annihilation factor is:

$$c_a(\lambda) = a_s(\lambda) + b_s(\lambda) \left[m^{-1}\right] \qquad (2.50)$$

Fermat's principle, also called the principle of least time, states that optical rays of light travel to the path of stationary optical length with respect to variations of the path, meaning that rays take the path that requires the least travel time. Clean air has the refraction index $n_2 = 1$ but the air into underground mine environments is usually filled with tiny suspended solid particles within the air that continuously move [125].

Light beam attenuation or transmission loss indicates that the intensity of the light emitted by LED increases with distance and the density of the medium.

The attenuation coefficients through the medium (in units of dB/m) are:

$$A = 10 * log_{10} \frac{I_{input}}{I_{output}} \quad (dB) \qquad (2.51)$$

In reality, the underground mining environment has lots of drawbacks when we analyze it from the optical channel behavior point of view. Therefore, a reliable, robust, and efficient VLC setup is very difficult to be realized due to a high number of variables and the complex conditions underground.

Both the position of the LED (embedded into the oTx) related to the position of the PD (embedded into the oRx), and the distance between them, can be easy determined for an ideal environment with clean air and a line of sight (LOS) topology with a short distance between LED and PIN PD for a proper, and robust VLC setup.

To study and model the optical channel impulse response (CIR), both IOP and AOP characteristics have to be taken in consideration in underground mine environments. Root mean square spread delay (RMS-DS) will not be considered in the underground mine environment for the CIR since the optical path considered is for a LoS topology setup with high optical attenuation (due to light's absorption and scattering).

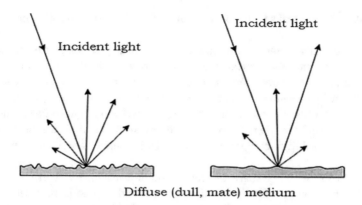

Diffuse (dull, mate) medium

Fig. 2.30 Light's behavior incident to surfaces and matter into underground spaces

The object's position (on the path of optical beam—between LED and PD), their geometry, colors, types of materials they are made of (e.g., wood, steel, fabric, or painted objects) and the kind of surfaces of the object (matt or glossy, smooth, or rough) inside the main galleries of working faces are also very important environmental characteristics of the optical channel from the AOPs point of view.

Since all the characteristics of IOP and AOP in industrial environments cannot be measured or properly calculated based on a general, comprehensive mathematical model, a proper estimation is possible to be done based on data acquisition during a wireless transmission based on visible light.

Figure 2.30 shows the light behavior incident to surfaces (walls, ceiling, floor, objects underground) and suspension particles in the air that describes underground mining optical channels.

Reflectance in IR is higher than that in visible light bands. The reflectance of different materials (that changes with the bands) is measured with spectrophotometers. Reflectivity of plastic walls is lower than plaster walls, floor, and ceiling in the main bands of visible light range. The reflectance of these materials is between 0.1 and 0.8. Reflectivity of the plastic walls is near 550 nm wavelength. From the beginning of 400 nm, as wavelength increases, reflectance of plaster walls, floor, and ceiling may grow slowly [109].

As discussed above, when the underground optical channel model is intended to be accurate described and determined, not only the geometry of the objects indoor, the material the objects are made of, their color and texture but also their diffuse or specular response of the light beam on the path between oTx and its pair oRx, have to be taken into consideration.

Another important aspect when the optical power at the receiver is calculated and the CIR mathematical model is estimated, the polluted air with different suspended particles of coal and rocks, has to be carefully taken into consideration, as well.

There are two important subjects from the CIR point of view. Ceiling and walls' color and their roughness (material composition), different objects inside (conveyors and other mining equipment and materials) as well as the polluted air with tiny suspended solid particles of dust (coal and rocks) in motion, close to the working face.

Since the underground coal mining environment is predominant black, without sunlight source and low optical power from other artificial sources of light, the AWGN has low value and therefore interferences from natural light are absent and from other artificial sources are almost negligible. One more advantage stands in the dark shade of walls, floor as well as ceiling that will stop light to have reflections from multiple surfaces (there is no multipath propagation and ISI can be neglected) and therefore the diffuse component of the general channel modeling formula can be neglected.

Close to the working space (where the mining machinery work), the optical channel is filled with a moving mixture of coal and tiny solid particles of stone that makes it difficult to identify the CIR model as close as possible to reality. Unlike the research on free optical space outdoors where the shape of raindrops and fog is considered, being almost the same with a regular spherical shape, the shape of coal and rock particles is irregular. Since coal particles cannot be treated simply as spheres the light ray's behavior (scattering) is quite difficult to be estimated [126]. The diameters of the smallest pores in coal are about 0.5–1 nm [127].

Changing the communication distances, evaluation of BER can be done. As long as the distance between oTx and oRx is decreased, the BER increases, also. The reason is that, as long as the distance between oTx and oRx increases, the light intensity becomes weaker, it scatters, the optical signal is attenuated and therefore the oRx cannot make the difference between high/low level of light, resulting in high BER. In case of a Rayleigh scattering, the propagation optical loss through environment is typically high due to the path loss, being proportional to $1 = 1/\lambda^4$.

2.2.4   Modulation Techniques for VLC Setup

Data communication in optical wireless transmission is possible through IM/DD of the incoherent light sources and therefore, the transmitted signals have to be positive with real values [39].

The oTx driver circuit in a VLC setup controls the current flowing to the LED and the brightness of the light radiated is modulated, therefore, the signal modulated in VLC setups consists of light pulses, being limited by the LED. The 3-dB modulation bandwidth of the currently commercially available LEDs rises to MHz order (2–20 MHz [39, 128]).

To increase the throughput, beside the improved both LED structure (with a shorter time of the "rise and fall") and oTx driver design, the most appropriate modulation technique has to be correctly embedded in the VLC's architecture.

In OWC setups, optical power/light intensity is subject to eye safety rules and design requirements and constraints, therefore, all the modulation techniques developed have to follow the regulations imposed by BS EN 62471:2008 [129].

Modulation schemes used in visible light wireless data communication are classified into two categories:

- single-carrier modulation techniques (SCMT),
- multi-carrier modulation techniques (MCMT).

Modulation schemes applied for wireless data in single carrier communication are achieved by altering the periodic waveform or the carrier, consisting of frequency, amplitude, or phase, therefore, there are three basic single carrier techniques to convert a digital sequence into a pulse one: changing the amplitude, position, or width of the pulse.

SCMTs such as OOK and PPM are embedded with low complexity circuits, but the most important drawback is that at high frequencies they cannot overcome the multipath effects resulting in ISI. More efficient and advanced modulation have been recently applied in order to overcome the LED's limited modulation bandwidth and insufficient data rate transmission.

Both for high data rates and ISI mitigation, though, many advanced modulations such as OFDM, color shift keying (CSK), CAP, DMT, or Nyquist single carrier (N-SC) have been proposed [130]. MCMTs such as OFDM, CAP, or DMT [131] have higher spectral efficiency and are able to overcome multipath effects [132].

Single-Carrier Modulation Techniques (SCMTs)

In 1997, Kahn and Barry were the first who proposed SCMT for IM/DD applied for IR wireless communication [41].

The general diagram of the IM/DD optical communication system, focused on the visible light area, is presented in Fig. 2.31.

In the electrical domain (ED), data are processed by the electrical modulator (EM) and then by the optical intensity modulator (OIM). Signal is represented here by an electrical voltage or current ($s(t)$). In either case (voltage or current), the electrical power is proportional to $s(t)^2$. The OIM generates an optical signal with intensity of $s_i(t)$. The optical power is proportional with $s(t)$. The signal $s(t)$ can have only real and positive values, therefore, the modulation techniques usually applied in

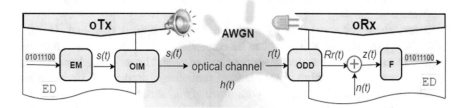

Fig. 2.31 IM/DD general diagram for VLC. *ED* electrical domain, *EM* electrical modulator, *OIM* optical intensity modulator, *AWGN* Additive White Gaussian Noise, *ODD* optical direct detection, *F* matched filter

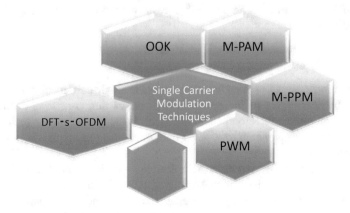

Fig. 2.32 Single-carrier modulation techniques. *OOK* On–off keying, *M-PAM* multilevel pulse amplitude modulation, *M-PPM* multilevel pulse position modulation, *PWM* pulse width modulation, *CAP* carrierless amplitude modulation, *DFT-s-OFDM* discrete Fourier transformation spread OFDM

radio communication have to be modified. On the optical channel with impulse response $h(t)$, an additive white Gaussian noise (AWGN) will be added and interfere with the optical signal sent by the LED to the PD. Here, the signal is represented by the optical intensity.

The optical direct detection (ODD) consists of the active area of the PD, converting the signal from the optical to electrical form, therefore the signal

$$r(t) = h(t) \otimes s(t) \tag{2.52}$$

will become an electrical one again $Rr(t)$ (voltage or current).

In the electrical domain (*ED*), the PD's noise signal

$$z(t) = Rr(t) \otimes n(t) \tag{2.53}$$

where:

$n(t)$—shot noise and thermal noise.

A matched filter (*F*) detection is the last stage before the final data out is completed.

Today, there are commercially available low-cost optical front-end devices, LEDs, and PDs used in VLC that can be embedded as IM/DD systems. Both the amplitude and phase of the electromagnetic waves cannot be modulated or detected by LEDs and PDs, therefore, most classical modulation used in RF transmission signals can be applied in VLC only with modified/enhanced techniques. Since off-the-shelf LEDs emit incoherent light, they are able to convey data wirelessly by the intensity of the light signal [133].

Common single carrier modulation techniques (SCMT) (Fig. 2.32.) used for VLC setups are OOK, multilevel pulse amplitude modulation (M-PAM), Multilevel pulse

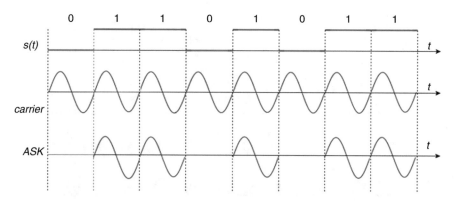

Fig. 2.33 Waveform of ASK modulation signal

position modulation (M-PPM) and PWM, discrete Fourier transformation spread OFDM (DFT-s-OFDM), and CAP [134], most of them already used in IR communications, as well.

At high data rates, RMS DS of the channel becomes similar to or even larger than the pulse duration, when SCMTs are applied in VLC setups, resulting in severe ISI and thus low BER performance and SNR penalty [39].

OOK also known as binary amplitude shift keying (ASK), uses an unipolar not return to zero (NRZ) code sequence to control the opening and closing of a sinusoidal carrier. This modulation considers the optical pulse that spreads to the entire part of the 1 bit duration. Hence, the presence of carrier is considered to be the binary 1 and its absence, the binary 0. Both phase and frequency of the carrier are constant during ASK modulation. It is the simplest modulation technique for IM/DD in VLC architectures.

OOK is a good trade-off between performance and complexity, considering that off-the-shelf hardware can be easily implemented with this modulation technique. Different more complicated schemes consider different duration in order to transmit additional data. This technique is an analog to unipolar encoding. Although it is easy to implement, there are several issues regarding illumination control and data throughput.

The waveform of the ASK is mathematically represented as:

$$s(t) = m(t) \cdot \sin\left(2\pi \cdot f_c \cdot t\right) \tag{2.54}$$

where:
  $s(t)$—the ASK output signal.
  $m(t)$—unipolar binary message signal that has to be transmitted.
  f_c—carrier's frequency.
  t—time.

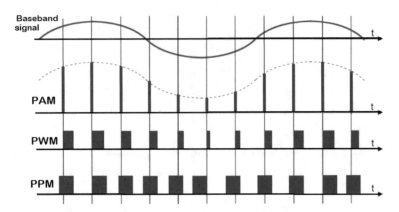

Fig. 2.34 PAM, PWM, and PPM modulation

ASK is a narrowband digital modulation scheme where the amplitude of the sinusoid carrier is modified according to the value of the modulated symbol (Fig. 2.33).

OOK is considered today the simplest modulation technique that can use the Manchester encoding for DC balance for data communication in free optical space.

In 2009, both Oxford University and South Korea's Samsung applied the OOK NRZ and achieved 100 Mbps data transmission at a low BER. Various practical implementation have been also presented with different equalization techniques [38, 41, 135].

Li et al. designed in 2014 a VLC system that achieved 340 Mbps using a post equalization circuit with OOL NRZ modulation [136].

PAM, PWM, and PPM are modulation methods that encode data transmitted in the amplitude (PAM), width (PWM), and position (PPM) of the pulse.

PAM consists of varying the amplitude of the waveform pulse corresponding to the signal variation mode. It can be observed (Fig. 2.34) that each sample of the strobed signal, modulates the amplitude of a pulse.

PWM uses a constant amplitude pulse train with determined successive time intervals, varying the width according to the modulator signal samples (Fig. 2.34).

PPM uses a series of pulses of constant duration and amplitude as carrier, while the position of the pulse relative to the sampling moments is variable (Fig. 2.34).

PPM, compared to OOK, has higher power efficiency and signal bandwidth. In PPM, m message bits are encoded with one single pulse in one of 2^m possible time shifts. This technique takes t seconds, therefore, the transmitted bit rate is m/t bits per second. This modulation technique is applicable in VLC setups for environments with zero or just few multipath interference. PPM is widely used and has high error performance, good power efficiency, moderate bandwidth efficiency, and the pulse duration very short.

Multilevel PPM has been developed in order to increase the PPM's spectral efficiency by conveying multiple pulses per symbol time, having the potential to achieve high spectral efficiency [137].

M-PPM (data encoding in the position of the pulse) has been considered for IR communication by Audeh et al. [138] due to its robust, low complexity, and low SNR through indoor channel.

PWM encodes the message into a pulsing signal. Several bits of data can be carried within each pulse.

M-PAM technique is used to modulate the incoming bits according to the amplitude of the optical pulse. In PPM case, the position of the optical pulse is modulated into shorter duration with a position that fluctuates, depending on the incoming bits [134].

OFDM has a higher optical power efficiency than OOK or PPM due to its resistance to ISI. Classical OFDM is widely used in conventional wired and wireless broadband communication systems, where signals are bipolar and complex. OFDM for optical communication, IM/DD, has few constraints and thus, different improved techniques.

The well-known practical implementation of OFDM consists of using inverse fast Fourier transform (IFFT) operation applied in oTx on a block of symbols from M-ary quadrature amplitude modulation (M-QAM), one of the conventional digital modulation techniques.

This technique relies on the mapping of different M-QAM symbols on the time domain signal into subcarriers/bands in the frequency domain, which unfortunately cannot be applied straight in VLC since it generates complex-valued samples, and IM/DD requires real nonnegative signals.

QAM as a combination of amplitude modulation (AM) and PSK is a system in which data are transmitted by modulating the amplitude of two separate carriers out of phase by 90° (cosine and sine). Due to their difference in phase, they are named quadrature carriers. The original stream of data is split into two sequences that consist of even and odd symbols.

By imposing a Hermitian symmetry constraint in IFFT operation [139], real but bipolar samples will result. In order to generate an unipolar OFDM signal, many techniques have already been developed. One of them, commonly used, is DCO-OFDM that allows to introduce a positive DC bias level, around which the bipolar signal can be applied. Since the main disadvantage of this technique relies on the increased energy dissipation of the oTx, different alternative methods with a significant energy enhancement have been developed (see Fig. 2.35).

The OFDM's important features are data transmitted in parallel with different sinusoidal subcarriers (having different frequencies) and the cyclic prefix (CP). OFDM is a spectrum efficiency modulation technique being resilient to both ISI and narrow-band effects.

All the subcarriers are jointly orthogonal over each symbol period. A general representation of both modulation and demodulation stages is presented in Fig. 2.36.

The serial input data are partitioned in blocks. Each block is mapped in a vector

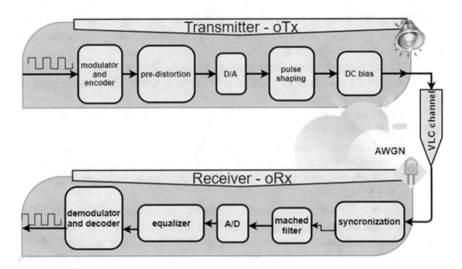

Fig. 2.35 PPM and PAM modulations. *D/A* Digital to Analog conversion, *A/D* Analog to Digital conversion

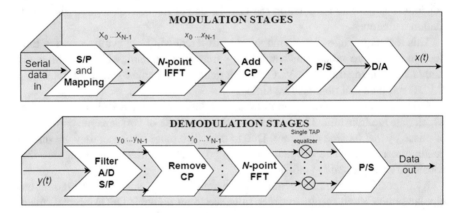

Fig. 2.36 General representation of modulation and demodulation stages for VLC. *S/P* serial/parallel, *IFFT* inverse fast Fourier transform, *CP* cyclic prefix, *D/A* digital/analog, *A/D* analog/digital, *FFT* fast Fourier transform, *P/S* parallel/serial

$$(X_{0,i} \ldots X_{N-1,i}) \text{ (in frequency domain)}$$

with the length of N complex numbers.

The complex numbers represent the QAM constellation.

The IFFT operations are applied to obtain parallel OFDM symbols and discrete time representation as follows:

$$x[k] = \sum_{n=0}^{N-1} X_n e^{j2\pi nk/N} \ , \quad k = 0, 1, \ldots, N-1 \tag{2.55}$$

where:

X_n—symbol sent at the nth subcarrier.

The vector $X_{0,i} \ldots X_{N-1,i}$ is constrained to have Hermitian symmetry meaning $X_n = X^*_{N-n}$. These complex numbers are constellation points according to modulation technique, being used for each subcarrier, from 4-QAM to 64-QAM, most frequently 16-QAM. Although PSK is compatible with OFDM, it is rarely used in VLC systems, because PSK in OFDM (unlike PSK in single carrier systems), does not have a constant signal envelope. Therefore, in case of large constellations, with smaller distance between points in constellation, PSK in OFDM is more vulnerable to noise [140].

The sequence of the complex numbers resulting from the constellation mapping is S/P converted to the vector $X_{0,i} \ldots X_{N-1,i}$ suitable for input into the IFFT stage [141].

A N-point IFFT of $X_{0,i} \ldots X_{N-1,i}$ generates the vector with complex numbers $x_{0,i} \ldots x_{N-1,i}$ (in time domain).

For a large number of subcarriers, $x[k]$ has a Gaussian distribution:

$$\sigma^2 = 2\left(\frac{N}{2} - 1\right) \tag{2.56}$$

The OFDM symbol rate R_s, for a given bit rate R_b, is:

$$R_s = \frac{2R_b}{\log M} \tag{2.57}$$

Adding the cycle prefix (CP), to eliminate the ISI at oRx, the sampling rate of the OFDM becomes:

$$f_s = r_{os} R_s \frac{N + N_{CP}}{N} \tag{2.58}$$

where:

r_{os}—oversampling ratio

N_{CP}—length of CP

Oversampling ratio (r_{os}) related with the number of used subcarriers, is expressed as:

$$r_{os} = \frac{N}{N_u} \tag{2.59}$$

In VLC systems, the CP has to be longer than the delay spread of the optical channel and the oRx must have an accurate synchronization with oTx. These conditions eliminate ISI and guarantee that the received subcarriers are orthogonal over the useful symbol period. Data are converted from serial to parallel after the addition of CP, then is converted to analog representation and finally is filtered to generate a continuous signal $x(t)$ (in time domain) which is real not complex due to Hermitian symmetry applied. The OFDM symbol has the bandwidth (BW) defined as $f_s/(2r_{os})$.

After the CP has been added and serial to parallel conversion was final, the signal $x[k]$ has to be clipped resulting in the clipped signal $x_c[k]$.

The clipped signal (1) follows the constraint of IM/DD to be strictly positive, (2) passes the nonlinearities of the LED, and (3) follows the dynamic range of the D/A converter as well as the A/D converter.

For the clipping ratio of r_1 and r_2, the clipped signal becomes $x_c[k]$ with the two levels—σr_1 and σr_2.

According to Elgala et al., the LED's behavior into the non-clipping region is linear [142].

Considering $H_{DAC}(f)$, the frequency response of D/A converter, with the cut-off frequency being equal to BW of the OFDM' symbols. According to Perin et al., low-pass frequency is modeled by a fifth-order Bessel filter [140].

Following the D/A conversion, the DC bias response is:

$$H_{DC\_bias} = 2r \sum_{n=1}^{\frac{N}{2}-1} \left| H_{\frac{D}{A}}(f_n) \right|^2 \qquad (2.60)$$

where:

$H_{D/A}$—digital to analog converter response

f_n—the frequency of the nth subcarrier

The LED's frequency response ($H_{LED}(f)$) with cut-off frequency f_{cLED} is modeled by Chen et al. as a low-pass filter with first-order Butterworth filter [142].

The optical signal is sent by LED, over the free channel with the CIR $h(t)$ having a frequency response of $H_{CIR}(f)$. At the oRx, both the PD's and TIA's frequency response of an antialiasing filter in the AC/DC, $H_{A/D}(f)$. As in the D/A converter, the A/D cutoff frequency is set equal to the BW of the OFDM' symbols [143].

This limitation reduces to N/2 from N, the number of independent values per symbol sent. For a 4-QAM means that per OFDM symbol, 2 bits can be sent.

Most of the stages in the demodulator are the same as in the modulator, except the single TAP equalizer. The main benefit of multipath transmission in OFDM with CP is that it has frequency nonselective fading on each individual subcarrier, so the magnitude and phase of each subcarrier are changed but ISI or inter-carrier interference (ICI) are absent. The single TAP equalizer will correct the phase and magnitude of each subcarrier, carrying out a single complex multiplication per each subcarrier [144].

CAP modulation consists of two orthogonal signals modulated similar with QAM and in phase (Fig. 2.37.) without overhead and carrier [145].

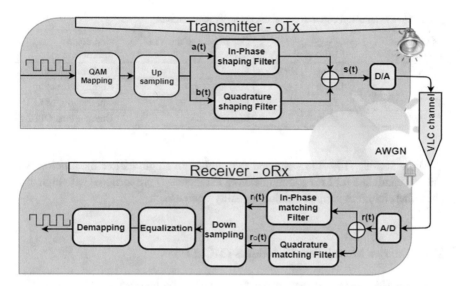

Fig. 2.37 CAP modulation diagram (adapted from [130]). Following up sampling, $a(t)$ and $b(t)$ are the original bit sequences. Following digital to analog conversion, $s(t)$ are the transmitted bit sequences. *AWGN* additive white Gaussian noise, Following analog to digital conversion, $r(t)$ is the received bit sequence. Following in-phase matching filter, $r_I(t)$ is the bit sequence. Following quadrature matching filter, $r_Q(t)$ is the bit sequence

CAP increases the optical wireless communication capacity and has high spectral efficiency. The input data are mapped by an encoder into two independent multilevel symbol streams. Each of the two streams passes through in-phase filter and quadrature shaping filter, whose IR forms a Hilbert transform pair, making both streams orthogonal.

The orthogonal streams are added and passed to a D/A converter, and as in OFDM technique, a DC bias is added in order to result in an unipolar signal. At the oRx, the inverse Hilbert filters, and then a decision feedback equalizer (DFE) extracts the symbol streams [146].

Compared with OFDM, CAP modulation is simpler and has lower PAPR, since the two orthogonal signals don't use IFFT and FFT blocks. On the other hand, since the equalizer is quite complex, it negatively affects the entire system's performance [146].

Table 2.5. displays a comparison between some single carrier modulations presented above.

Using a finite impulse response (FIR) filter, Haigh et al., proposed the CAP modulation, splitting the CAP spectrum in m-subcarriers [147] (see Fig. 2.36.).

In the case of a DFT-s-OFDM, the modulated waves were communicated separately using multiple LEDs in a single array and have been demonstrated with low PAPR. Taking into consideration both DFT-s-OFDM and DCO-OFDM, Wu et al. demonstrated a higher performance, reported to PAPR and BER, of DFT-s-OFDM [148].

Table 2.5 Comparison between some single carrier modulations

Modulation technique	Power Efficiency	Spectral	Complexity of the VLC setup	Additional information
OOK	Low	High	Low	Flickering possibility
PPM	High	Low	Medium	Complex structure
PAM	Low	Medium	Low	Nonlinearity of LED
CAP	High	High	Medium	Cheaper than OFDM

Many studies [149–153], showed that, although single carrier modulation is a suitable candidate for LiFi communication, ISI increases proportional with high data rates, therefore, the communication's quality decreases.

Multi-Carrier Modulation Techniques (MCMTs)

Islim and Hass [134], in their paper "*Modulation Techniques for LiFi*" address a subdivision of MCMT based on OFDM as seen in Fig. 2.38.

Bipolar signals cannot be sent as IM/DD because, as the literature shows, the intensity of light cannot be negative, the values can be real and positive only, therefore an adapted OFDM has to be considered.

OFDM modulation technique encodes digital data on multiple carrier frequencies. Its key feature relies on the fact that data are parallel transmitted on a number of sinusoidal subcarriers of different frequencies, therefore closely signal spaced orthogonal subcarriers, carry data in parallel channels. The frequencies are chosen in a way that subcarriers are reciprocally orthogonal over each OFDM symbol period [154].

With this method, it is possible to modulate each carrier in the same bandwidth with a conventional modulation technique PSK or QAM at a low symbol rate.

In OFDM adapted for VLC, RMS DS is considerably shorter than the symbol duration. Using CP and a large number of subcarriers, OFDM technique excludes the possibility of ISI's appearance. The highest value of the channel's excess delay is smaller than the CP. In this way, the indoor channel becomes from dispersive, flat fading over the subcarrier bandwidth [155].

The CP has the main key feature to provide an additional guard interval to eliminate ISI from the previous symbol. CP repeats the end of the symbol, so, before each OFDM symbol is a copy of the end part of the same symbol.

Using a sampling rate of 20 MHz, two samples of CP can compensate the ISI of maximum DS up to 100 ns, thus, for the entire OFDM frame, for BDs up to 20 MHz, the indoor optical communication channel can be considered flat [41, 155]. A number of subcarriers equal or higher than 64 showed that the time-domain signal has a Gaussian distribution [156].

The most applied unipolar OFDM modulation techniques, both in VLC and LiFi setups, are DCO-OFDM [158] (Fig. 2.39.) and ACO-OFDM [159] (Fig. 2.40).

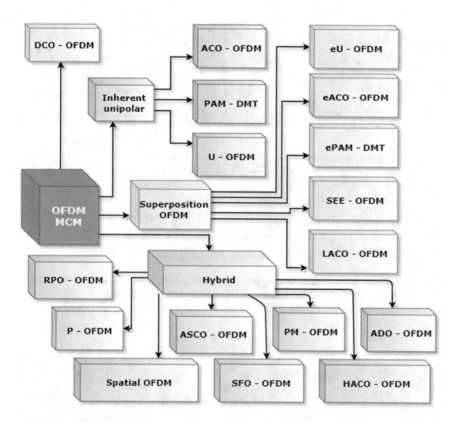

Fig. 2.38 The most used OFDM multicarriers modulations (adapted from [134]). *DCO-OFDM* DC biased OFDM, *ACO-OFDM* Asymmetrically Clipped Optical OFDM, *PAM-DMT* Pulse Amplitude Modulation—Discrete Multitone, *U-OFDM* Unipolar OFDM, *eU-OFDM* enhanced Unipolar OFDM, *eACO-OFDM* enhanced Asymmetrically Clipped Optical OFDM, *ePAM-DMT* enhanced PAM-DMT, *SEE-OFDM* Spectrally and Energy Efficient OFDM, *LACO-OFDM* Layered ACO-OFDM, *RPO-OFDM* Reverse Polarity Optical OFDM, *P-OFDM* Polar OFDM, *ASCO-OFDM* Asymmetrically and Symmetrically Clipped Optical OFDM, *SFO-OFDM* Spectrally Factorized Optical OFDM, *PM-OFDM* Position Modulation OFDM, *HACO-OFDM* Hybrid Asymmetrically Clipped Optical OFDM, *ADO-OFDM* Asymmetrically Clipped DC biased Optical OFDM

In case of ACO-OFDM, $Xn = 0$ for n even. For a given R_b bit rate, the OFDM symbol rate (R_s) becomes:

$$R_s = 2p\frac{R_b}{log_2M} \tag{2.61}$$

where:

M—constellation size (considering that is the same for all subcarriers).

$p = 1$ for DC-OFDM.

$p = 2$ for ACO-OFDM.

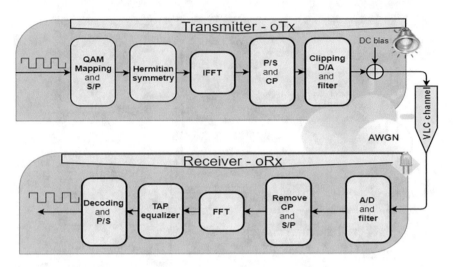

Fig. 2.39 Diagram of DCO-OFDM (adapted from [130]). *QAM* quadrature amplitude modulation, *S/P* serial/parallel, *IFFT* inverse fast Fourier transformation, *P/S* parallel/serial, *CP* cyclic prefix, *D/A* digital/analog, *A/D* analog/digital, *FFT* fast Fourier transformation

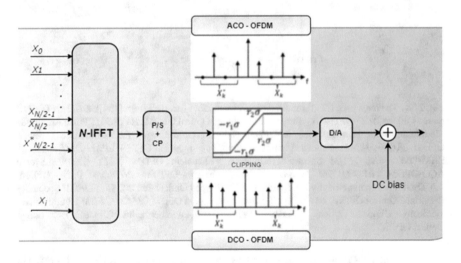

Fig. 2.40 Comparison on stages between DCO-OFDM and ACO-OFDM

In order to preserve the bit rate, following CP, the actual sampling rate of the OFDM (f_s) must be increased:

$$f_s = \frac{N + N_{CP}}{N} r_{os} R_s \tag{2.62}$$

where:

N_{CP}—length of CP
r_{os}—oversampling ratio
N—number of subcarriers used.
R_s—OFDM symbol rate.
The number of subcarriers used to carry data is:

$$N_u = \frac{N}{pr_{os}} \tag{2.63}$$

In order to reduce the necessary dynamic range of the D/A, after P/S conversion and insertion of CP, the discrete-time OFDM signal $x(k)$ is clipped at levels—$r_1\sigma$ and $r_2\sigma$.

$$x_c(k) = \begin{cases} -r_1\sigma, & for \ x(k) \leq -r_1\sigma \\ x(k), & for \ r_1\sigma < x(k) < \ r_2\sigma \\ r_2\sigma, & for \ x(k) \geq r_2\sigma \end{cases} \tag{2.64}$$

where:
$r_1 = r_2 = r$ for DC-OFDM.
$r_1 = 0$, and $r_2 = r$ for ACO-OFDM.
r_1 and r_2 are referred to as clipping ratios.
The clipping probability (P_c):

$$P_c = Q(r_1) + Q(r_2) \tag{2.65}$$

where:
$Q(r_i)$ is the Q-function for the tail probability of a Gaussian distribution [140].
D/A converts the clipped signal into analog domain and DC bias is added to make the signal positive.
In case that both A/D and an electrical to optical conversion have unit DC gain, the average optical power is equal to electrical signal with the DC bias added.
In case of DC-OFDM, the average optical power is:

$$\overline{P} = r\sigma \tag{2.66}$$

In case of ACO-OFDM, the average optical power is:

$$\overline{P} = \frac{\sigma}{\sqrt{2\pi}} \tag{2.67}$$

where:
σ—the variance of the signal samples distribution.

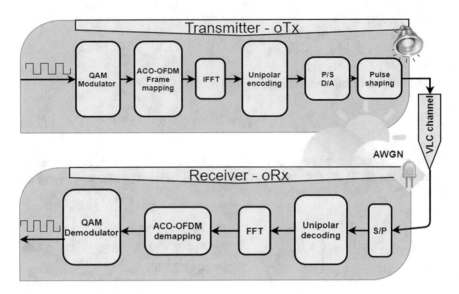

Fig. 2.41 Diagram of ACO-OFDM (Adapted from [157]). *QAM* quadrature amplitude modulation, *IFFT* inverse fast Fourier transformation, *P/S* parallel/serial, *D/A* Digital/Analogue, *FFT* Fast Fourier Transformation

In DCO-OFDM modulation technique, all the subcarriers are modulated and a DC-bias is added to the signal to make it positive. When the DC-bias has a high value, the PAPR becomes high, leading to a low power efficiency.

Salma D. Mohamed et al. showed that, in case of ACO-OFDM, the even subcarriers are set to zero and therefore data are mapped to the odd subcarriers only. Moreover, the output negative parts from IFFT are clipped to zero so, the odd subcarriers are not affected since the clipping noise falls on even subcarriers. This technique decreases to half the data rate, hence has half data spectral efficiency compared to DCO-OFDM. However, ACO-OFDM is more power efficient than DCO-OFDM since it does not include DC-bias [159].

Inherent unipolar OFDM (Fig. 2.41) refers to asymmetrically clipped optical OFDM (ACO-OFDM) proposed by Armstrong and *Lowery* in 2006 [157], pulse amplitude modulation discrete multitone (PAM-DMT) [160], and unipolar OFDM (U-OFDM) [161], such as ACO-OFDM [160].

In LiFi systems, where high illumination level is required, Hadamard coded modulation (HCM) was proposed for multicarrier modulation. This modulation is based on fast Walsh–Hadamard transform (FWHT) proposed as an alternative to the FFT.

Compared to ACO-OFDM and DCO-OFDM at high illumination levels, HCM achieves higher performance gains, according to Noshad et al. [162].

DC reduced HCM (DCR-HCM) was proposed as an alternative to HCM, both to support dimmable LiFi applications and reduce its power consumption.

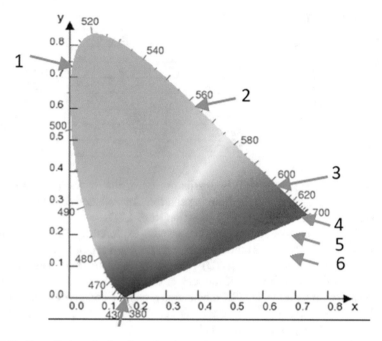

Fig. 2.42 Constellation of colors from the chromatic CIE 1931

Changing LED's color is a simple technique comparable to carrier frequency modulation. RGB-LEDs are able to illuminate with different colors depending on the light intensity applied to each of the LED's elements.

The IEEE 802.15.7/2011 standard defines CSK as modulation technique for VLC [107]. CSK uses multi-chip LEDs for VL. The received bits are mapped into constellation points of colors according to chromatic CIE 1931 color space. CIE 1931 color spaces define the links, in quantitative terms, between the physiologically perceived colors in human color vision and distributions of wavelengths in the visible spectrum, as shown in Fig. 2.41.

The CIE1931 is the illumination model for the human eye color perception (Fig. 2.42). Any color in this model can be represented by its chromaticity dimension [x, y]. In CSK modulation technique, the general intensity of the output color is constant, thus the instantaneous color of the RGB LED is modulated. The relative intensities between the multiple colors used in a model are, however, changed [163].

The CSK modulation technique works by maintaining both a constant illumination color and optical power while the light intensity of each element of the RGB LED is constantly changed to match the necessary constellation point. Normally a constant illumination color avoids LEDs flickering.

The symbol mapping of 6 CSK on the CIE 1931 color model based on IEEE 802.15.7 is presented in Table 2.6.

Even though the center of the CIE 1931 color constellation is kept constant, the amplitude dimming is possible by controlling the LED' brightness in CSK

Table 2.6 CIE1931 color model

Position	Code	Band (nm)	Center (nm)	(x,y)
0	000	380–478	429	(0.169, 0.007)
1	001	478–540	509	(0.011, 0.733)
2	010	540–588	564	(0.402, 0.597)
3	110	588–633	611	(0.669, 0.331)
4	100	633–679	656	(0.729, 0.271)
5	101	679–726	703	(0.734, 0.265)
6	110	726–780	753	(0.734, 0.265)

modulation. Still, the color shift is possible to occur due to the presence of any improper driving current used for dimming control. Constellation sizes up to 16 CSK were presented in the IEEE 802.15.7 based on RGB LEDs.

The CIE 1931 constellation points design was explored by Drost et al. using billiard algorithms [163], by Monterio et al. with the interior point method [164], by Singh et al. using quad LED (QLED) [165], and by Jiang et al. using extrinsic transfer charts for an iterative CSK transceiver design [166].

Practical implementation of CSK modulation requires a complex circuit structure. A feedback loop at oRx can be used for color calibration to avoid interference from other natural or artificial light sources [167].

2.2.5   IEEE Standard for VLC

IEEE approved and published in 2011 the standard 802.15.7 for VLC. It defines the short range WOC using visible light and supports data rates up to 96 Mbps by fast modulation of the optical signal received from light sources which are possible to be dimmed. It also provides dimming adaptable mechanisms to avoid LED's flickering at high data rate for visible light data communication purpose [168].

The standard defines also the physical layer (PHY) possible to be applied in outdoor applications (even though its speed of 267.6 Kbps is quite slow) and medium access control (MAC) layer.

PHY layer is divided into three types—PHY I, II, and III—each of them describing a combination of different modulation schemes:

- PHY 1 layer relates to outdoor application and works from 11.67 to 267.6 Kbps.
- PHY 2 layer extents data rates from 1.25 to 96 Mbps.
- PHY 3 is used for many optical sources with the modulation technique CSK delivering data rates between 12 and 96 Mbps [167].

CSK recommended in IEEE 802.15.7 to improve the data rates that were rather low in different other modulation schemes. The switching capacity slows down by generating white light using blue LEDs with yellow phosphor. As a result, an alternate procedure to obtain the white light is the use of three separate RGB

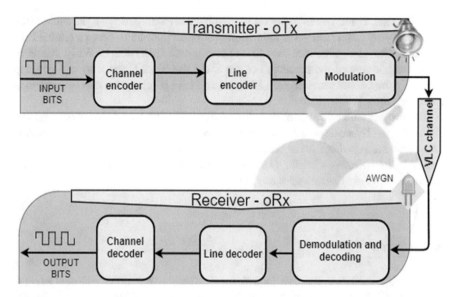

Fig. 2.43 PHY layer for VLC

LEDs. Modulation in CSK is accomplished using the RGB LED. CSK depends on the color space chromaticity diagram (Fig. 2.41.). It charts all colors perceivable by the human eye related to two chromaticity parameters x and y [169].

The modulation formats recognized for layers PHY I and II are OOK and variable PPM (VPPM). The Manchester coding used for layers PHY I and II takes into account the clock representing a logic 0 with an OOK symbol "01" and a logic 1 with an OOK "10," all with a DC component that eludes light fading in case of an extended presence of logic 0's.

Figure 2.43 presents a typical PHY layer system model of VLC.

MAC layer handles PHY layer management issues related to addressing, collision avoidance, and data acknowledgment protocols and also supports three multiple access topologies: peer-to-peer, star configuration, and broadcast.

The MAC layer allows using the link with the other layers as transmission control protocol/internet protocol (TCP/IP) protocol.

The tasks performed by the MAC layer include mobility, dimming, visibility, and security support. It also includes schemes for flickering mitigation, color function support, network beacon generation if the device is a coordinator, variable VPAM disassociation, and association support, providing a consistent link between MAC entities.

This standard delivers data rates in order to send video, audio as well as multimedia services. It concerns optical transmission mobility, its compatibility with artificial lighting present in already installed infrastructure of lighting fixtures, as well as the interference generated by the ambient lighting.

IEEE Task Force works today on the Global LiFi standard for light while 2019 marks the year when LiFi gains international support in order to adopt a Global LiFi standard. The new guidelines for LiFi are to improve IEEE 802.11 Wi-Fi standard. IEEE has announced the formation of the 802.11bb Task Force who will develop and ratify the Global standard for LiFi, ensuring LiFi is integrated into every device and every light [170].

The project under development is the "*IEEE 802.11bb Standard for Information Technology—Telecommunications and Information Exchange Between Systems Local and Metropolitan Area Networks—Specific Requirements, Part 11: Wireless LAN Medium Access Control (MAC) and Physical Layer (PHY) Specifications Amendment: Light Communications*" [128].

References

1. [Online] https://www.cisco.com. https://www.cisco.com/c/en/us/solutions/collateral /service-provider/visual-networking-index-vni/mobile-white-paper-c11–520862.html. Last accessed 28 March, 2020.
2. Aldridge, D. (1973). *Analysis of communication systems in coal mines*, West Virginia University, Morgantown, West Virginia, U.S.A., USBM Grant Final Report G0101702 (MIN–39), May 1.
3. Bandyopadhyay, L. K., Chaulya, S. K., Mishra, P. K. (2010). *Wireless Communication in Underground Mines, RFID-Based Sensor Networking*, Springer Science Business Media, LLC, ISBN 978–0–387-98164-2.
4. Olsen, R. G., & Farstad, A. J. (1973). Electromagnetic direction finding experiments for location of trapped miners. *IEEE Transactions on Geoscience Electronics, 11*(4), 178–185.
5. Durkin, J. (1984). Electromagnetic detection of trapped miners. *IEEE Communication Magazine, 22*(2), 37–46.
6. Kravitz, J. H., Kovac, J. G. and Duerr, W. H. (1994). *Advances in mine emergency communication*. Proceedings of the International Emergency Management and Engineering Conference, Hollywood Beach, Florida, pp. 23–26.
7. Kumar, A., Chaulya, S. K., Kumar, S., & Bandyopadhyay, L. K. (2004). Trapped miners detection, location and communication system. *Minetech, 24*(6), 3–13.
8. Vong, P. K., Lai, H. C., & Rodger, D. (2006). Modeling electromagnetic field propagation in eddy-current regions of low conductivity. *IEEE Transactions on Magnetics, 42*(4), 1267–1270.
9. Higginson, M. (1992). *Radio Propagation Experiments Durban Deep Goldmine. Draft report on investigation*, COMRO, September 1992, 4 p.
10. Boutin, M., Benzakour, A., Despins, C. L., & Affes, S. (2008). Radio wave characterization and modeling in underground mine tunnels. *IEEE Transactions on Antennas and Propagation, 56*(2), 540–549.
11. Wait, J. R. (1989) *Electromagnetic and electrochemical response of geological conductors*. Proceedings of IEEE International Symposium of Antennas and Propagation Society, California, USA, 2, 26–30 June 1989, pp. 1156–1159.
12. Austin, B. A. (1978). *Underground radio communication techniques and system in South African mines*, Proceedings of the Workshop on Electromagnetic Guided Waves in Mine Environment, Boulder, CO, Institute for Telecommunication Science, 28–30 March 1978, pp. 87–102.

13. Durkin, J. (1982). *Performance evaluation of electromagnetic techniques for location of trapped miners.* Report of Investigations 87II, US Bureau of Mines.
14. Kononov, V. A. and Higginson, M. R. (1994). *Trapped miner locator (marketing feasibility),* CSIR Miningtek, Final report Y5I62, Johannesburg, South Africa.
15. [Online] https://www.csir.co.za/technologies-improve-safety-mines, Last accessed 28 March 2020.
16. Nessler, N. H. (2000). Electromagnetic location system for trapped miners. *Subsurface Sensing Technologies and Applications, 1*(2), 229–246. https://doi.org/10.1023/A:1010172810336.
17. Sydanheimo, L., Keskilammi, M., Kivikoski, M. (2000). *Reliable mobile computing to underground mine.* Proceedings of IEEE International Conference on Communications, Louisiana, USA, Vol. 2, 18–22 June 2000, pp. 882–888.
18. Vasquez, J., Rodriguez, V., & Reagor, D. (2004). Underground wireless communications using high-temperature superconducting receivers. *IEEE Transactions on Applied Superconductivity, 14*(1), 46–53.
19. Srinivasan, K., Ndoh, M. and Kaluri, K. (2005). *Advanced wireless networks for underground mine communications.* Proceedings of First International Workshop on Wireless Communications in Underground and Confined Areas, Val-d'Or, Québec, Canada, 6–7 June 2005, pp. 51–54.
20. Akyildiz, I. F., & Stuntebeck, E. P. (2006). Wireless underground sensor networks: Research challenges. *Ad Hoc Networks, 4*(6), 669–686.
21. Yarkan, S., Guzelgoz, S., Arslan, H., & Murphy, R. R. (2009). Underground mine communications: A survey. *IEEE Communications Surveys & Tutorials, 11*(3), 125–142. 3rd Quarter.
22. Kennedy, G. A., & Bedford, M. D. (2014). *Underground wireless networking: A performance evaluation of communication standards for tunnelling and mining. Tunnelling and Underground Space Technology Volume, 43,* 157–170.
23. Nüchter, A., Elseberg, J., Borrman, D. (2013). *Irma3D—an intelligent robot for mapping applications.* In: 3rd IFAC Symposium on Telematics Applications – Proceedings, Seoul, South Korea, pp. 119–124.
24. Reddy, A. H., Kalyan, B., & Murthy Ch, S. N. (2015). Mine rescue robot system—A review. *Procedia Earth and Planetary Science, 11,* 457–462. https://doi.org/10.1016/j.proeps.2015.06.045.
25. Rosca, S., Riurean, S., Leba, M., & Ionica, A. (2019). A reliable wireless communication system for hazardous environments. In T. Antipova & A. Rocha (Eds.), *Digital science. DSIC18 2018. Advances in intelligent systems and computing* (Vol. 850). Cham: Springer.
26. Hu, B., Wang, Z. (2014). *A cross-layer congestion control algorithm for underground video transmission over wireless networks.* In: 11th IEEE International Conference on Networking, Sensing and Control—Proceedings, Miami, FL, USA, pp. 239–244.
27. Sunderman, C. and Waynert, J. (2012). *An overview of underground coal miner electronic tracking system technologies,* IEEE Industry Applications Society Annual Meeting, Las Vegas, NV, 2012, pp. 1–5.
28. Hedley, M., Gipps, I. (2013), *Accurate wireless tracking for underground mining,* IEEE International Conference on Communications Workshops, ICC2013—Proceedings, Budapest, Hungary, pp. 42–46.
29. Dayekh, S., Affes, S., Kandil, N., Nerguizian, C. (2014). *Cost-effective localization in underground mines using new SIMO/MIMO-like fingerprints and artificial neural networks,* IEEE International Conference on Communications Workshops, ICC 2014—Proceedings, Sydney, NSW, Australia, pp. 730–735.
30. Chehri, A., Farjow, W., Mouftah, H., & Fernando, X. (2009). *Design of wireless sensor network for mine safety monitoring, 24th Canadian conference on electrical and computer engineering (CCECE)* (pp. 1532–1555). Canada: Niagara Falls.

31. Forooshani, A., Bashir, M., Michelson, D., Noghanian, S. (2013). *A survey of wireless communications and propagation modeling in underground mines*. IEEE Commun. Surv. Tutorials 15, 1524–1545, 4th Quarter.

32. Dohare, Y., Maity, T., Paul, P., Das, P. (2014). *Design of surveillance and safety system for underground coal mines based on low power WSN*, International Conference on Signal Propagation and Computer Technology (ICSPCT), Ajmer, India, pp. 116–119.

33. Dohare, Y., Maity, T., Das, P., & Paul, P. (2015). Wireless communication and environment monitoring in underground coal mines—Review. *IETE Technical Review, 32,* 140–150.

34. Paavola, I. M., Seppälä, P. (2016). *Wireless networks in underground mines, Industrial Wireless Sensor Networks,* Monitoring, Control and Automation. Wood head Publishing Series in Electronic and Optical Materials, Pages 107–123, DOI: https://doi.org/10.1016/B978-1-78242-230-3.00006-4.

35. Zhou, C., & Jacksha, R. (2016). Modeling and measurement of radio propagation in tunnel environments. *IEEE Antennas and Wireless Propagation Letters, 16,* 1431–1434. https://doi.org/10.1109/LAWP.2016.2561903.

36. Ndoh, M. (2004). *Modélisation de la propagation des ondes électromagnétiques dans un environnement minier,* Thèse de Ph. D., Département de génie électrique et génie informatique, Université de Laval, Québec, Canada.

37. Borah, D. K., Boucouvalas, A. C., Davis, C. C., Hranilovic, S., & Yiannopoulos, K. (2012). A review of communication-oriented optical wireless systems. *EURASIP Journal on Wireless Communications and Networking, 1,* 91.

38. Sarpeshkar, R., Delbruck, T., & Mead, C. A. (1993). White noise in MOS transistors and resistors. *IEEE Circuits and Devices Magazine, 9*(6), 23–29. https://doi.org/10.1109/101.261888.

39. Dimitrov, S. and Haas, H. (2015). Principles of LED light communications. Towards *Networked LiFi,* Cambridge. Cambridge University Press.

40. Bapst, R. F., & Gfeller, U. (1979). Wireless in-house data communication via diffuse infrared radiation. *Proceedings of the IEEE, 67*(11), 1474–1486.

41. Kahn, R., & Barry, J. M. (1997). Wireless infrared communications. *Proceedings of the IEEE, 85*(2), 265–298.

42. Meyer-Arendt, J. R. (1968). Radiometry and photometry: Units and conversion factors. *Applied Optics, 7*(10), 2081–2084. https://doi.org/10.1364/AO.7.002081.

43. Collins, S., O'Brien, D. C., & Watt, A. (2014). High gain, wide field of view concentrator for optical communications. *Optics Letters, 7*(39), 1756–1759.

44. Manousiadis, P. P., Rajbhandari, S., Mulyawan, R., Vithanage, H., Chun, G., Faulkner, D. C., O'Brien, G. A., Turnbull, S., Collins, I. D., & Samuel, W. (2016). Wide field-of-view fluorescent antenna for visible light communications beyond the etendue limit. *Optica, 3*(7), 702–7016. https://doi.org/10.1364/AO.7.002081.

45. Marsh, M., & Kahn, J. (1997). Channel reuse strategies for indoor infrared wireless communications. *IEEE Transactions on Communications, 45,* 1280–1290.

46. Barry, J. R., Kahn, J. M., Krause, W. J., Lee, E. A., & Messerschmitt, D. G. (1993). Simulation of multipath impulse response for indoor wireless optical channels. *IEEE Journal on Selected Areas in Communications, 11*(3), 367–379.

47. Pakravan, M. R., Kavehrad, M., & Hashemi, H. (2001). Indoor wireless infrared channel characterization by measurements. *IEEE Transactions on Vehicular Technology, 50,* 1053–1073.

48. [Online] https://www.nordiclights.com/techtalk/phenom-optics/, Last accessed 28 March, 2020.

49. [Online] Nobel Prize https://www.nobelprize.org/nobel_prizes /physics/laureates /2014/ popular-physics prize2014.pdf. (2014). Last accessed 28 March, 2020.

50. Oxlade, C. (2012). *Tales of Inventions. The light bulb.* London: Capston Global Library Ltd.

51. Novikov, M. A. and Losev, O. V., *Pioneer of Semiconductor Electronics.* Physics of the Solid State, Translated from Fizika Tverdogo Tela, No. 1, pp. 5–9, (2004).

52. Virk, H. S. (2015). History of luminescence from ancient to modern times. *Defect and Diffusion Forum, 361*, 1–13.
53. Ramirez-Iniguez, R., Idrus, S. M., & Sun, Z. (2008). *Optical wireless communications: IR for wireless connectivity*. CRC Press.
54. Lupei, V. and Lupei, A. (2015). *Nd:YAG at its 50th anniversary: Still to learn,* Journal of Luminescence, DOI: https://doi.org/10.1016/j.jlumin.2015.04.018.
55. Ghassemlooy, Z., Popoola, W., & Rajbhandari, S. (2013). *Optical wireless communications, system and channel modelling with MATLAB*. CRC Press, Tylor and Francis Group.
56. Haas, H., Chen, C., & O'Brian, D. (2017). A guide to wireless networking by light. *Progress in Quantum Electronics, 55*, 88–111.
57. Tsonev, D., Sinanović, S. and Haas, H. (2012). *Novel Unipolar Orthogonal Frequency Division Multiplexing (U-OFDM) for Optical Wireless*. Yokohama, Japan: IEEE, May 6–9, Vol. Proceedings of the Vehicular Technology Conference (VTC Spring).
58. Kwon, D. H., Yang, S. H. and Han, S. K. (2015). *Modulation bandwidth enhancement of white-LED-based visible light communications using electrical equalizations*, San Francisco, California, United States SPIE OPTO, Vols. Proc. SPIE 9387, Broadband Access Communication Technologies IX, 938. DOI: https://doi.org/10.1117/12.2078680.
59. [Online] https://www.lumileds.com/uploads/377/WP17-pdf. Last accessed 28 March, 2020.
60. Will, P. A., & Reineke, S. (2019). *Organic light-emitting diodes, handbook of organic materials for electronic and photonic devices* (2nd ed., pp. 695–726). Woodhead Publishing Series in Electronic and Optical Materials. https://doi.org/10.1016/b978-0-08-102284-9. 00021-8.
61. [Online] https://www.sldlaser.com/why-laserlight, Last accessed 28 March, 2020.
62. Chi, Y., Hsieh, D., Lin, C., et al. (2015). Phosphorous diffuser diverged blue laser diode for indoor lighting and communication. *Scientific Reports, 5*, 18690. https://doi.org/10.1038/srep18690.
63. Chi, Y., Huang, Y., Wu, T., et al. (2017). Violet laser diode enables lighting communication. *Scientific Reports, 7*, 10469. https://doi.org/10.1038/s41598-017-11186-0.
64. Huang, Y., Chi, Y., Kao, H., et al. (2017). Blue laser diode based free-space optical data transmission elevated to 18 Gbps over 16 m. *Scientific Reports, 7*, 10478. https://doi.org/10. 1038/s41598-017-10289-y.
65. Bahanshal, S., Alwazani, H., Majid, M. A. (2019). *Design of RGB Laser Diode Drivers for Smart Lighting and Li-Fi using MATLAB GUI*, Conference (2019) 1st International Conference on Electrical, Control and Instrumentation Engineering (ICECIE), November doi: https://doi.org/10.1109/ICECIE47765.2019.8974772.
66. Lee, C., Sufyan, I. M., Videv, S., Sparks, A., Shah, B., Rudy, P., McLaurin, M., Haas, H., Raring, J. (2020). *Advanced LiFi technology: Laser light*, Proceedings Volume 11302, Light-Emitting Devices, Materials, and Applications XXIV; 1130213 (2020) doi:https://doi.org/10. 1117/12.2537420, SPIE OPTO, San Francisco, California, U.S.
67. [Online] https://physics.nist.gov/cgi-bin/cuu/Value?e. Last accessed 28 March, 2020.
68. [Online] The NIST Reference on Constants, Units and Uncertainity. https://physics.nist.gov / cgi-bin/cuu/Value?klsearch_for=Boltzmann %E2%80%99s+ constant+. Last accessed 28 March, 2020.
69. Juan-de-Dios, S.-L., Arvizu, A., Mendieta Francisco, J., & Nieto, H. I. (2011). Advanced trends in wireless Communications. *IntechOpen*. https://doi.org/10.5772/15493.
70. Le, H. M., O'Brien, D., Faulkner, G., Lubin, Z., Kyungwoo, L., Daekwang, J., & YunJe, O. (2008). High-speed visible light communications using multiple-resonant equalization. *IEEE Photonics Technology Letters, 20*, 1243–1245.
71. Randel, S., Breyer, F., Lee, S. C. J., & Walewski, J. W. (2010). Advanced modulation schemes for short-range optical communications. *IEEE Journal of Selected Topics in Quantum Electronics, 20*, 1280–1289.

72. Vucic, J., Kottke, C., Nerreter, S., Buttner, A., Langer, K. D., & Walewski, J. W. (2009). White light wireless transmission at 200+ Mb/s net data rate by use of discrete-multitone modulation. *IEEE Photonics Technology Letters, 21*, 1511–1513.

73. Chitnis, D., Zhang, L., Chun, H., Rajbhandari, S., Faulkner, G., O'Brien, D. and Collins, S. A. (2016). *200 Mb/s VLC demonstration with a SPAD based receiver*, s.l.: IEEE Summer Topicals Meeting Series (SUM). pp. 226–227.

74. Li, Y., Safari, M., Henderson, R., & Haas, H. (2015). *Optical OFDM with single-photon avalanche diode. IEEE Photonics Technology, 27*(9), 943–946. https://doi.org/10.1109/LPT.2015.2402151.

75. Sarbazi, E. and Haas, H. (2015). *Detection statistics and error performance of SPAD-based optical receivers,* IEEE 26th Annual International Symposium on Personal, Indoor, and Mobile Radio Communications (PIMRC), 2015, pp. 830–834, IEEE 26th Annual International Symposium on Personal, Indoor, and Mobile Radio Communications (PIMRC), pp. 830–834, DOI: https://doi.org/10.1109/PIMRC.2015.7343412.

76. [Online] *Thorlabs*, "*Bandpass Filters*". https://www.thorlabs.com search/thorsearch.cfm?search=bandpass%20filter. Last accessed28 March, 2020.

77. Mulyawana, R., Gomeza, A., Chuna, H., Rajbhandarib, S. and Manousiadisc, P. P. (2017). *A comparative study of optical concentrators for visible light communications*, et.al. SPIE OPTO., IEEE Photonics Technology Letters, pp. 99–105, doi: https://doi.org/10.1117/12.2252355.

78. Collins, S., O'Brien, D. C., & Watt, A. (2014). High gain, wide field of view concentrator for optical communications. *Optics Letters, 39*, 20. https://doi.org/10.1364/OL.39.0017562014.

79. Chi, N. (2018). *LED-based visible light communications. Signals and Communication Technology*, Tsinghua University Press, Beijing doi: https://doi.org/10.1007/978-3-662-56660-2.

80. [Online] Electronics-tutorials. https://www.electronics-tutorials.ws/ transistor/ tran_8.html. Last accessed 28 March, 2020.

81. Balogh, L.., www.ti.com. Fundamentals of MOSFET and IGBT Gate Driver Circuits. Last accessed 28 March, 2020 [Online].

82. Toumazou, C., Lidgey, F. J. and Haigh, D. G. (1990). *Analogue IC Design: The Current Mode Approach,* s.l.: IEE Circuits, Devices and Systems Series, Peter Peregrinus Ltd., USA.

83. Fujimoto, N. and Mochizuki, H., *477 Mbit/s visible light transmission based on OOK-NRZ modulation using a single commercially available visible LED and a practical LED driver with a pre-emphasis circuit.* [ed.] OSA Publishing. Anaheim, CA, USA, Optical Fiber Communication Conference, pp. 1–3, DOI: https://doi.org/10.1364/NFOEC.2013.JTh2A.73, (2013).

84. Li, H., Chen, X., Guo, J., & Chen, H. (2014). A 550 Mbit/s real-time visible light communication system based on phosphorescent white light LED for practical high-speed low-complexity application. *Optics Express, OSA Publishing, 22*(22), 27203–27213. https://doi.org/10.1364/OE.22.027203.

85. Haigh, P., Bausi, F., Ghassemlooy, Z., Papakonstantinou, I., Minh, H., Le, F. C., & Cacialli, F. (2014). *Visible light communications: real time 10 Mb/s link with a low bandwidth polymer light-emitting diode. Optics Express, 22*, 2830–2838. https://doi.org/10.1364/OE.22.002830.

86. Huang, X., et al. (2015). *750Mbit/s visible light communications employing 64QAM-OFDM based on amplitude equalization circuit.* LoS Angeles, CA, USA, Proceedings of the Optical Fiber Communications Conference and Exhibition. pp. 1–3.

87. Zhou, Y. et al., *2.08Gbit/s visible light communication utilizing power exponential pre-equalization.* s.l.: 25th Wireless and Optical Communication Conference (WOCC), pp. 1–3, DOI: https://doi.org/10.1109/WOCC.2016.7506539, (2016).

88. Uysal, M., Capsoni, C., Ghassemlooy, Z., Boucouvalas, A. and Udvary, E. (2016). *Optical Wireless Communications. An Emerging Technology,* Springer International Publishing Switzerland, ISBN 978–3–319-30201-0.

89. Chan, Y.-J., Chien, F.-T., Shin, T.-T. and Ho, W.-J. (2002). *Bandwidth enhancement of transimpedance amplifier by capacitive peaking design, 6353366* U.S.A.

90. Morikuni, J. J., & Kang, S.-M. (1992). An analysis of inductive peaking in Photoreceiver design. *IEEE Journal of Lightwave Technology, 10*(10), 1426–1437.

91. Shekhar, S., Walling, J. S., & Allstot, D. J. (2006). Bandwidth extension techniques for CMOS amplifiers. *IEEE Journal of Solid-State Circuits, 41*(11), 2424–2439.

92. Figueiredo, M., Alves, L. N., & Ribeiro, C. (2017). Lighting the wireless world: The promise and challenges of visible Light communication. *IEEE Consumer Electronics Magazine, 4*, 28–37. https://doi.org/10.1109/mce.2017.2714721.

93. Böcker, A., Eklind, V., Hansson, D., Holgersson, F., Nolkrantz, J.and Severinson, A. (2015). *An implementation of a Visible Light Communication system based on LEDs,* Gothenburg, Sweden, Chalmers University of Technology. Department of Signals and Systems Division of Communication Systems.

94. Ghassemlooy, Z., Alves, L. N., Zvánovec, S., & Khalighi, M.-A. (2017). *Visible Light Communications. Theory and applications.* CRC Press Taylor & Francis Group.

95. Alves, L. N. and Aguiar, R. L., *Design techniques for high performance optical wireless front-ends,* Aveiro, Portugal, Proceedings of the Conference on Telecommunications—ConfTele, 2003.

96. Razavi, B. (2012). *Design of integrated circuits for optical communications.* New York: McGraw-Hill.

97. Sindhubala, K. and Vijayalakshmi, B., *Simulation of VLC system under the influence of optical background noise using filtering technique.* Issue 2, Part B, s.l. Science Direct, Elsevier, Materials Today: Proceedings, Volume 4, pp. 4239–4250 (2017).

98. Adiono, T., Putra, R. V. W., & Fuada, S. (2018). Noise and bandwidth consideration in designing op-amp based transimpedance amplifier for VLC. *Bulletin of Electrical Engineering and Informatics, 7*, 312–320. https://doi.org/10.11591/eei.v7i2.870. 2, June 2018.

99. Chang, F-L., Hu, W-W, Lee, D. and Yu, C (2017). *Design and implementation of anti low-frequency noise in visible light communications.* [ed.] International Conference on Applied System Innovation (ICASI). Sapporo. pp. 1536–1538, DOI:https://doi.org/10.1109/ICASI.2017.7988219.

100. Karimi-Bidhendi, A., Mohammadnezhad, H., Green, M. M., & Heydari, P. (2018). A silicon-based low-power broadband Transimpedance amplifier. *IEEE Transactions on Circuits and Systems I: Regular Papers, 65*(2), 498–509. https://doi.org/10.1109/TCSI.2017.273.

101. Lee, S. J., Kwon, J. K., Jung, S. Y. and Kwon, Y. H. (2012). *Simulation modeling of visible light communication channel for automotive applications,* Anchorage, USA, Sept. 2012, Proc. IEEE ITSC'12, pp. 463–468, DOI: https://doi.org/10.1109/ITSC.2012.6338610.

102. Sarbazi, E., Uysal, M., Abdallah, M. and Qaraqe, K. (2014). *Ray tracing based channel modeling for visible light communications.* Trabzon, Turkey, Proc. SPCA'14, pp. 23–25.

103. Miramirkhani, F., & Uysal, M. (2015). Channel modelling and characterization for visible light communications. *IEEE Photonics Journal, 7*(6), 1–16. https://doi.org/10.1109/JPHOT. 2015.2504238.

104. Al-Kinani, A., Wang, C.-X., Haas, H., and Yang, Y. (2016). *A geometry-based multiple bounce model for visible light communication channels,* Paphos, Cyprus: in Proc. IEEE IWCMC'16, pp. 31–37.

105. Miramirkhani, F., Narmanlioglu, O., Uysal, M., & Panayirci, E. (2017). A mobile channel model for VLC and application to adaptive system design. *IEEE Communications Letters, 21* (5), 1035–1038.

106. [Online] *Git Hub* https://github.com/mhrex/ Indoor_VLC_Ray_Tracing. Last accessed 28 March, 2020.

107. [Online] IEEE Standard for Local and Metropolitan Area Networks, Part 15.7:, *Short- Range Wireless Optical Communication Using Visible Light.* IEEE Std. 802.15.7· 2011, (2011).

108. Ramirez-Aguilera, A. M., Luna-Rivera, J. M., Guerra, V., Rabadan, J., Perez-Jimenez, R., & Lopez-Hernandez, F. J. (2018). *A generalized multi-wavelength propagation model for VLC indoor channels using Monte Carlo simulation.* John Wiley & Sons Ltd, Trans Emerging Tel Tech. https://doi.org/10.1002/ett.3490.

109. Yang, Q., Chen, H.-H. and Meng, W.-X (2016). *Channel modeling for visible light communications—a survey*, Ed. Wirel. Commun. Mob. Computer, Wiley Online Library, Wireless Communications and Mobile Computing, pp. 2016–2034.
110. Carruthers, J. B., & Kannan, J. M. (2002). Iterative site-based modeling for wireless infrared channels. *IEEE Transactions on Antennas and Propagation, 50*(5), 759–765.
111. Carruthers, J. B., & Kahn, J. M. (1997). Modeling of nondirected wireless infrared channels. *IEEE Transactions on Communications, 4510*, 1260–1268.
112. Lopez-Hernandez, F. J., & Betancor, M. J. (1997). DUSTIN: Algorithm for calculation of impulse response on IR wireless indoor channels. *IEEE Electronic Letters, 33*(21), 1804–1806.
113. Al-Kinani, A., Wang, C.-X., Zhou, L., & Zhang, W. (2018). Optical wireless communication channel measurements and models. *IEEE Communications Surveys and Tutorials*. https://doi.org/10.1109/COMST.2018.2838096.
114. Jungnickel, V., Pohl, V., Nonnig, S., & Helmolt, C. V. (2002). A physical model of the wireless infrared communication channel. *IEEE Journal on Selected Areas in Communications, 20*(3), 631–640.
115. Ding, J., Wang, K., and Xu, Z. (2014). *Accuracy analysis of different modelling schemes in indoor visible light communications with distributed array sources*, in Proc. IEEE CSNDSP'14, Manchester, UK, July 2014, pp. 1005–1010.
116. Lopez-Hernandez, F. J., Perez-Jimenez, R., & Santamaria, A. (1998). Monte Carlo calculation of impulse response on diffuse IR wireless indoor channels. *IEEE Electronic Letters, 34*(12), 1260–1262.
117. Haas, H. *Wireless data from every light bulb*, http://bit.ly/tedvlc. TED Website, August (2011) [Online].
118. Rappaport, T. S. (2001). *Wireless communications: Principles and practice,* 2nd Edition. [ed.] Prentice Hall., Prentice Hall Communications Engineering and Emerging Technologies, Dec 31.
119. Goldsmith, A. (2005). *Wireless communications,* s.l.: Cambridge University Press.
120. Jivkova, M., & Kavehrad, S. T. (2000). Multispot diffusing configuration for wireless infrared access. *IEEE Transactions on Communications, 48*(6), 970–978.
121. Viswanathan, M. (2020). *Gaussian waves.* https://www.gaussianwaves. com/2014/07/power-delay-profile/. 9 July (2014). Last accessed 28 March [Online].
122. Zeng, L., Brien, D. O., Le-Minh, H., Lee, K., Jung, D. and Oh, Y. (2008). *Improvement of data rate by using equalization in an indoor visible light communication system.* Shanghai Proceedings of the International Conference on Circuits and Systems for Communications, pp. 678–682.
123. Ijaz, M., Ghassemlooy, Z., Pesek, J., Fiser, O., LeMinh, H., & Bentley, E. (2013). Modeling of fog and smoke attenuation in free space optical communications link under controlled laboratory conditions. *Journal of Lightwave Technology, 31*, 11.
124. Wang, W.-Z., Yan-MingWang, G.-Q.S., and Wang, D.-M. *Numerical study on infrared optical property of diffuse coal particles in mine fully mechanized working combined with CFD Method*, Hindawi Publishing Corporation Mathematical Problems in Engineering Volume 2015, Article ID 501401, 10 pages doi:https://doi.org/10.1155/2015/501401.
125. McCartney, J. T., Ergun, S. (1968). Optical properties of coals and graphite, Dept of the Int. Bureau of Mines.
126. Cannon, H. C. G. and George, W., *Refractive Index of Coals*, 09 Jan. 1943, Publisher Nature 151 (1943).
127. Speight, J. G. (1994). *The chemistry and technology of coal.* New York: Marcel Decker.
128. Khalid, A. M., Cossu, G., Corsini, R., Choudhury, P., & Ciaramella, E. (2012). 1Gbit/s visible light communication link based on phosphorescent white LED. *IEEE Photonics Switching, 4*, 2. https://doi.org/10.1109/JPHOT.2012.2210397.
129. [Online] BS EN 62471:2008., *Photobiological Safety of Lamps and Lamp Systems*, BSI British Standards Std., Sept. (2008).

130. Ji, R., Wang, S., Liu, Q., & Lu, W. (2018). High-speed visible light communications: Enabling technologies and state of the art. *Applied Sciences, 8*(4), 589. https://doi.org/10.3390/app8040589.

131. Langer, K.-D. (2015). DMT modulation for VLC. In S. Arnon (Ed.), *Visible Light communication*. Cambridge: University Printing House.

132. Chi, N. (2018). *LED-based visible light communications. Signals and communication technology*. Beijing: Tsinghua University Press. https://doi.org/10.1007/978-3-662-56660-2.

133. Burchardt, H., Serafimovski, N., Tsonev, D., Videv, S., & Haas, H. (2014). VLC: Beyond point-to-point communication. *IEEE Communication Magazine, 52*(7), 98–105.

134. Islim, M. S. and Haas, H. (2016). http://www.cnki.net/kcms/detail/34.1294. TN.20160413.1658.002.html, Published online. DOI: 10.3969/j. ISSN.16735188.

135. Hou, R., Chen, Y., Wu, J. and Zhang, H. (2015). *A brief survey of optical wireless communication*. Sydney, Australia, Proceedings Australian Symposium on Parallel and Distributed Computing.

136. Li, H., Chen, X., Huang, B., Tang, D., & Chen, H. (2014). High bandwidth visible light communication based on a post-equalization circuit. *IEEE Photonics Technology Letters, 26*(2), 119–122.

137. Noshad, N. and Brandt-Pearce, M. (2013). *Can visible light communications provide Gb/s service?* https://arxiv.org/abs/1308.3217

138. Audeh, M. D., Kahn, J. M., & Barry, J. R. (1996). Performance of pulse-position modulation on measured non-directed indoor infrared channels. *IEEE Transactions on Communications, 44*(6), 654–659.

139. Wang, Z., Tsonev, D., Videv, S., & Haas, H. (2015). Unlocking spectral efficiency in intensity modulation and direct detection systems. *IEEE Journal on Selected Areas in Communications, 33*, 9. https://doi.org/10.1109/JSAC.2015.2432530.

140. Perin, J. K., Sharif, M., & Kahn, J. M. (2015). *Modulation schemes for single-laser 100 Gb/s links: Multicarrier,*. No. 24, s.l. *Journal of Lightwave Technology, 33*, 5122–5132.

141. Armstrong, J. (2009). *OFDM for optical communications*. 3, s.l. *Journal of Lightwave Technology, 27*, 189–204. https://doi.org/10.1109/jlt.2008.2010061.

142. Elgala, H., Mesleh, R., & Haas, H. (2009). *Predistortion in optical wireless transmission using OFDM*, Shenyang, China, s.n. *Ninth International Conference on Hybrid Intelligent Systems, 2*, 184–189.

143. Chen, C., Basnayaka, D. A., & Haas, H. (2016). Downlink performance of optical attocell networks. *Journal of Lightwave Technology, 34*, 137–156. https://doi.org/10.1109/JLT.2015.2511015. no. 1.

144. Armstrong, J., Schmidt, B. D. C , Kalra, D., Suraweera, H. A. and Lowery, A. J. (2006). *Performance of asymmetrically clipped optical OFDM in AWGN for an intensity modulated direct detection system*. San Francisco, CA, USA, 27 November - 1 December Proceedings of the Global Telecommunications Conference GLOBECOM '06. doi:https://doi.org/10.1109/GLOCOM.2006.571.

145. Wu, F.-M et al. (2013). *3.22-Gb/s WDM visible light communication of a single RGB LED employing carrier-less amplitude and phase modulation*. Anaheim, CA, USA, Proc. OFC/NFOEC, pp. 1–3.

146. Wu, F., et al. (2013). Performance comparison of OFDM signal and cap signal over high capacity RGB-led-based WDM visible light communication. *IEEE Photonics Journal, 5*(4), 7901507.

147. Haigh, P., Le, S. T., Zvanovec, S., et al. (2015). Multiband carrier-less amplitude and phase modulation for band limited visible light communications systems. *IEEE Wireless Communications, 22*(2), 46–53. https://doi.org/10.1109/MWC.2015.7096284.

148. Wu, C., Zhang, H. and Xu, W. (2014). *On visible light communication using LED array with DFT-spread OFDM*, Sydney, Australia, Jun. 2014, in IEEE International Conference on Communications (ICC), pp. 3325–3330 doi: https://doi.org/10.1109/ICC.2014.6883834.

149. Armstrong, B. J., & Schmidt, J. C. (2008). Comparison of asymmetrically clipped optical OFDM and DC-biased optical OFDM in AWGN. *IEEE Communications Letters, 12*(5), 343–345. https://doi.org/10.1109/LCOMM.2008.080193.

150. Mesleh, R., Elgala, H., & Haas, H. (2011). On the performance of different OFDM based optical wireless communication systems. *IEEE/OSA Journal of Optical Communications and Networking, 3*(8), 620–628. https://doi.org/10.1364/JOCN.3.000620.

151. Barros, D., Wilson, S., & Kahn, J. (2012). Comparison of orthogonal frequency division multiplexing and pulse amplitude modulation in indoor optical wireless links. *IEEE Transactions on Communications, 60*(1), 153–163. https://doi.org/10.1109/TCOMM.2011.112311.1.

152. Dissanayake, J., & Armstrong, S. (2013). Comparison of ACO-OFDM,DCO-OFDM and ADO-OFDM in IM/DD systems. *Journal of Lightwave Technology, 31*(7), 1063–1072. https://doi.org/10.1109/JLT.2013.2241731.

153. Kashani, M. and Kavehrad, M. (2014). *On the performance of single and multi-carrier modulation schemes for indoor visible light communication systems,* Austin, USA, in IEEE Global Communications Conference (GLOBECOM), pp. 2084–2089, doi: https://doi.org/10.1109/GLOCOM.2014.703.

154. Armstrong, J., Brendon, J., Schmidt, C., Kalra, D., Suraweera, H. A., & Lowery, A. J. (2006). *SPC07-4: performance of asymmetrically clipped optical OFDM in AWGN for an intensity modulated direct detection system,* S.l. IEEE Globecom, 1–5.

155. Armstrong, J. (2009). OFDM for optical communications. *Journal of Lightwave Technology, 27*(3), 189–204.

156. Dardari, D., Tralli, V., & Vaccari, A. (2000). *A theoretical characterization of nonlinear distortion effects in OFDM systems,* s.l. IEEE Transactions on Communications, 10(48), 1755–1764.

157. Armstrong, J., & Lowery, A. J. (2006). Power efficient optical OFDM. *Electronic Letters, 42* (6), 370–372. https://doi.org/10.1049/el:20063636.

158. Carruthers, J. B., & Kahn, J. M. (1994). Multiple-subcarrier modulation for non directed wireless infrared communication. *Proceedings IEEE Global Telecommunication Conference (GLOBECOM), 2,* 1055–1059.

159. Mohamed, S. D., Khallaf, H. S., Shalaby, H., Andonovic, I. and Aly, M. H. (2013). *Two approaches for the modified asymmetrically clipped optical orthogonal frequency division multiplexing system,* Second International Japan-Egypt Conference on Electronics, Communications and Computers (JEC-ECC), doi:https://doi.org/10.1109/jec-ecc.2013.6766400.

160. Lee, S. C. J., Randel, S., Breyer, F., & Koonen, A. M. J. (2009). PAM-DMT for intensity-modulated and direct-detection optical communication systems. *IEEE Photonics Technology Letters, 21*(23), 1749–1751.

161. Islim, M., Tsonev, D. and Haas, H., *Spectrally enhanced PAM-DMT for IM/DD optical wireless communications,* Proc. IEEE 25th Int. Symp. Pers. Indoor and Mobile Radio Commun (PIMRC). Hong Kong, China, pp. 927–932, doi:https://doi.org/10.1109/PIMRC.2015.7343421 (2015).

162. Noshad, M., & Brandt-Pearce, M. (2016). Hadamard coded modulation for visible light communications. *IEEE Transactions on Communications, 99,* 1167–1175. https://doi.org/10.1109/TCOMM.2016.2520471.

163. Drost, R. and Sadler, B., *Constellation design for color-shif keying using billiards algorithms.* Miami, USA, IEEE GLOBECOM Workshops (GC Wkshps), pp.980–984, doi:https://doi.org/10.1109/GLOCOMW.2010.5700472, (2010).

164. Monteiro, E., & Hranilovic, S. (2014). Design and implementation of color-shift keying for visible light communications. *Journal of Lightwave Technology, 32,* 2053–2060. https://doi.org/10.1109/JLT.2014.2314358.

165. Singh, R., O'Farrell, T., & David, J. P. R. (2014). *An enhanced color shift keying modulation scheme for high-speed wireless visible light communications.* No.14. *Journal of Lightwave Technology, 32,* 2582–2592. https://doi.org/10.1109/JLT.2014.2328866.

166. Jiang, J., Zhang, R., & Hanzo, L. (2015). Analysis and design of three-stage concatenated color-shift keying. *IEEE Transactions on Vehicular Technology, 64*(11), 5126–5136. https://doi.org/10.1109/TVT.2014.2382875.
167. Roberts, R. D., Rajagopal, S. and Lim, S., *IEEE 802.15.7 physical layer summary,* Houston, TX, s.n., 2011 IEEE GLOBECOM Workshops (GC Wkshps). pp. 772–776.
168. Ullah, K. L. (2016). *Visible light communication: Applications, architecture, standardization and research challenges.* [book auth.] Elsevier Digital Communications and Networks. 18 July, https://doi.org/10.1016/j.dcan.2016.07.004.
169. Leba M., Riurean S. and A. Ionica. *LiFi—The path to a new way of communication, 2017 12th Iberian Conference on Information Systems and Technologies (CISTI)*, Lisbon, 2017, pp. 1–6.
170. https://purelifi.com/lifi-is-getting-a-global-standard/, Last accessed 28 March, 2020 [Online].

Chapter 3
A Hybrid Communication System for Mining Industry. From RequirementS' Analysis to Testing the Product

The major aim of this chapter is to define and develop an integrated, solid, low-cost, and effective system of local and remote data transmission, based on the visible light wireless communication technology and conventional Ethernet, for the underground mining industry.

This chapter guides the reader through a complex process with different stages of development, from idea to implementation of a functional, and reliable final product. The stages presented move forward from concept, through design, implementation, till the testing of a VLC prototype.

When the investigation on the state-of-the-art of the existing functional systems is completed, the identification of all system's requirements is compulsory to be done for a proper development. The main parts of the system, their functional role, and characteristics, as well as all the rules and restrictions imposed by the special environment underground, must be clearly identified and considered. The requirements and specifications already defined, allow to establish the overall system architecture, therefore, simulation of the system with the support of different advanced dedicated software helps to speed up the entire development process and keeps the costs low. Each unit of the system is developed and tested for its functionality and then, all units are integrated into the final system and further tests are performed to evaluate its functionality into the difficult, harsh environments as the underground mining spaces are.

3.1 State-oF-the-ARt and RequirementS' Analysis

The working conditions underground are rough to work in, first, because of the harsh environment (polluted air, high temperatures, restricted visibility with low illumination level, high noise, confined work spaces, etc.), the high risk of explosion and the continuous inherent change of the infrastructure due to the coal exploitation. That

© Springer Nature Switzerland AG 2021
S. M. Riurean et al., *Application of Visible Light Wireless Communication in Underground Mine*, https://doi.org/10.1007/978-3-030-61408-9_3

Table 3.1 Potential RF emitters in a US coal mine [1]

Frequency	Application	Comments
300–10,000 Hz	Personal emergency devices	Through-the-earth communications
70–500 kHz	Proximity detection devices	Audible and visual warning
300–800 kHz	Medium frequency radios	Voice, text
150–175 MHz	Leaky feeder systems VHF	Voice and low bandwidth data
400–410 MHz	Miner or asset-tracking systems	Radio frequency identification (RFID)
450–470 MHz	Leaky feeder systems UHF	Voice and low bandwidth data
490 MHz	Remote-operated continuous miner	Remote control of continuous miner
900 MHz	Active radio frequency identification (RFID) tags	RFID to track miner's location
900 MHz	Line-of-sight radios	Voice, text
900 MHz	Rescue robots	Robot control
2.4 GHz	Rescue robots	Video
2.4 GHz	Line-of-sight radios	Voice, text

is why designing a solid and viable data communication network for the underground mining industry, which provides high reliability, is a challenging task.

Before 2006, a very limited number of intentional RF emitters were used underground. Today, following many disasters underground with numerous casualties, the new policies worldwide in mining occupational healthy and safe requires mine operators to install wireless, or partially wireless advanced systems with smart devices to support human life and warn in due time about the environment underground in order to avoid any possible calamity.

Because there is the potential for electromagnetic interference (EMI) when undesired EM energy from another RF system interferes with the reception or processing of a desired signal, causing unacceptable performance degradation to other systems, in Table 3.1, are listed the potential RF emitters with own frequencies and applications in US coal mines [1].

Another source of EMI is noise, comprising a random electrical voltage can originate within a radio receiver or can have an external origin. Power lines, electrical equipment (motors), electronic equipment (remote-control devices), transformers, and electrical/mechanical switching devices can generate EM noise. Electrically powered machinery used in mining also produces strong, low-frequency noise when starting up or when the power demand switches from low to high (or vice versa). Lightning is also a source of noise. This EM noise is low frequency, and the propagation loss is so low that its noise contributions could come from anywhere in the entire world. Wires that run into the mine can carry lightning and other EM noise can be generated from outside the mine.

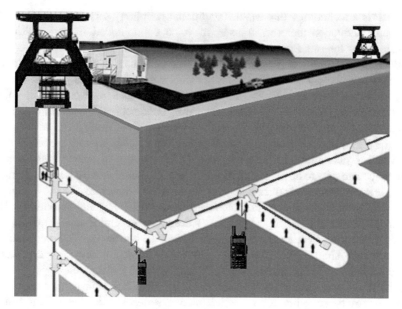

Fig. 3.1 A LFS system deployed underground (adapted from [2])

Different types of local and remote communication solutions, such as leaky feeder system (LFS), through-the-earth (TTE) transmission, Voice over Internet Protocol (VoIP), Wi-Fi, cellular network with Long Term Evolution (LTE), or radio system for short-range communication Ultra-Wideband (UWB), are now worldwide available for underground spaces in mines.

LFS—is a distributed antenna system using radiating coaxial cable installed along underground spaces that emits and receives radio waves (from 150 to 175 MHz for VHF and from 450 to 470 MHz for UHF). The cable has hollows or space shield on its outer conductor to allow the radio signal to leak out of the cable along its entire length and works as a very long antenna. Therefore, due to this signal leakage, line amplifiers are inserted at regular intervals (every 350–500 m), to increase the radiated signal to the optimal level. The signal is normally captured by portable radios worn on by workers. The transmissions of these portable devices are received by the radiating coaxial cable and are transported both to the surface and other parts of the tunnel, allowing a bidirectional communication [1].

An example of a LFS system deployed underground is presented in Fig. 3.1.

TTE transmissions are systems that use low-frequency electromagnetic waves to create a link between the surface and the underground mine. Communication is usually done by magnetic induction, with frequencies below 30 kHz. It has narrow bandwidth and large propagation loss. TTE systems are also susceptible to atmospheric noise and alternate-current harmonics from power lines and electrical equipment, which also represent deficiencies to a high-quality communication link, especially in the uplink connection [3].

VoIP is a technology that enables traditional telephony services to operate over computer networks, therefore transfer of voice communications and multimedia sessions over Internet Protocol (IP) networks is possible. VoIP technology enables traditional telephony services to operate over computer networks using packet-switched protocols. Packet-switched VoIP puts voice signals into packets that can be transmitted over any VoIP-compatible network, such as LAN. Some of the benefits of using VoIP underground are: it uses the existing IT network equipment and unshielded twisted pair (UTP) cabling to connect phone sets; it can be integrated with the traditional (digital or analogue) copper cable solutions with the use of voice gateways or analogue telephone adapter (ATA) devices; an integrated system for both voice and data to be managed. Nevertheless, the entire system is rather complex because it requires power over Ethernet (PoE) network equipment to provide power to devices such as phone sets. PoE (standardized by IEEE 802.3af—around 13 W of electrical power and 802.3 at—25.5 W) describes several standards or ad hoc systems that pass electric power along with data on twisted pair Ethernet cabling. This allows a single cable to provide both data connection and electric power to devices such as wireless access points, IP cameras, and VoIP phones.

The Wi-Fi wireless network (standardized by IEEE 802.11) has the main advantage to provide mobility for users but also a series of drawbacks: has limited coverage in the underground; requires site survey and planning prior to setting up; needs complex hardware to be configured; limited range resulting in high number of access points necessary to work properly; support a limited number of users since latency increases along with traffic; subject to interference; not applicable in sites with a high risk of explosion.

Cellular networking with LTE empowered. LTE refers to the fourth generation (4G) and beyond of cellular network. It is a high-speed wireless communication technology for mobile devices and data terminals based on IPs that provides transfer of voice, data, and video content with a better signal propagation compared to 2.4/5.0 GHz Wi-Fi in underground sites. Although it offers many benefits, such as high end to end Quality of Service (QoS) capabilities natively implemented in the standard, it also has important drawbacks, because it requires RF licensing for use in surface and underground installations or special protection in order to be approved to be used in mines with potentially explosive gases.

UWB (IEEE 802.15.4a standard, with a security extension specified in IEEE 802.15.4z) is a local wireless communication technology that uses radio waves with a wide spectrum of several GHz. It operates in LoS for a short range (up to 200 m) at a very high frequency. A UWB Tx works by sending billions of pulses across the wide spectrum frequency and a corresponding Rx translates the pulses into data by listening to a familiar pulse sequence that is sent by the Tx. UWB has high accuracy in real time since one pulse is sent every two nanoseconds. UWB is extremely low power, but its high bandwidth (500 MHz) is ideal to communicate data, from a host to other devices. To increase UWB's range and reception reliability, a system of MIMO antennas can be used to enable a reliable short-range network. The antennas can be embedded into any smart devices such a smartphone, a wristband, a belt, or a helmet. As important drawbacks, the UWB is solely a local communication technology, and has no capability of sending accurate remote data through walls.

A reliable and robust network system for local (in underground spaces) and remote (from underground to the surface) communication, consists mainly of two parts: the transceivers and the communication network. Both wireless and the wired (cable-based) communication systems for data transmissions and voice must be adjustable according to specific characteristics of the site especially because of the constant change of the infrastructure during exploitation.

Older technologies that supported voice communication, required separate networks. New technologies have integrated in many mines most of the old voice communication systems into the new data communication systems.

To keep the cost low, and simplify the entire communication system (both deployment and operations), one of the major objectives is to standardize transmissions and run multiple services along a single cabled backbone. To deploy a solid integrated communication system, even in case of a short-term planning of an optimal or at least near-optimal solutions for long-term exploitation, all technical and environmental conditions have to be taken into consideration.

The many types of communication systems can be classified due to the purpose of the network, into:

- tracking system of personnel and/or machinery,
- network for the emergency response in case of fatality,
- daily basis ventilation and ventilation on demand (VoD) in case of necessity,
- nonstop environmental monitoring and notification in case of emergency situation,
- a combination of the above.

Integrating all the above into a single communication system (simplex, duplex, or full-duplex) is a difficult task especially because of their different priority level of importance and speed of development.

Depending on the purpose of the network and the application(s) required, the proper communication system and its infrastructure, wireless (in RF or optical spectrum), or cabled (wired or optical fiber) or a hybrid system applies.

There are many requirements both for hardware and software parts of the entire system, especially when a full duplex communication architecture is planned to be designed. The more complex the requirements are, the more complex the communication system and its design, implementation, maintenance, and its expansion are.

All the characteristics of the sites and the various technologies must be considered in designing the network for all the spaces into the underground mine and from underground to surface when the proper applications, technologies, and communication infrastructure are selected.

The use of any of these solutions in underground spaces depends on several factors, the most important of all being the miners' safety due to the potentially dangerous situations during operation, mine's infrastructure, and its level of hazard risk both for workers and machinery.

A communication system for underground has to be designed, installed, and used taking into consideration the mine's topology. The mine's topology itself is also influenced by the type of mining, such as mineralization in the exploitation area and the surrounding areas; temperature and level of humidity underground; and the presence of various gasses.

These factors powerfully influence the mine structure, the topology of the communications network chosen, as well as the hardware used for communication (such as cables, enclosures, and dedicated network equipment).

Depending on the type of the rock exploited, hard rock (extraction of minerals such as gold, copper, or nickel) or soft rock (extraction of coal), the mine exploitations differentiate in distinct categories. With soft rock extraction, means of mechanical excavation can conduct mining with or without the use of explosives.

The network topologies have to be designed and developed to cover all open-spaces underground with many various and particular shapes with multilevel horizons.

The well-known communication network physical topologies (line, bus, tree, ring, star, extended star, partial mesh, or mesh) used as Ethernet networks indoor or outdoor, can be applied underground according to the specificity of the mine' topology.

Methods of exploitation (mainly differentiated by the walls and roof composition underground) can also influence the network solution to be implemented. The walls and roof can be supported when soft rock is exploited, or unsupported when no pillars are necessary. The access from surface to the underground workings can be done via a ramp, a vertical shaft (from the surface to multiple levels underground), or a combination of them (shaft(s) and underground ramps that connect two or more levels.

Each mining method outcomes different patterns of tunnels, resulting in specific topology for each mine. Regular mines comprise a series of shafts, ramps, drifts, and slopes across multiple levels designed to access the working faces where the minerals are. Given the hazardous and potentially destructive nature of mining (such as blasting and drilling), sensitive networking equipment, and cables are difficult to be installed and protected.

The shafts, ramps, drifts, and slopes across multiple levels in mine are the main paths used to transport personnel, minerals, materials, and equipment in and out of the mine. Along with this permanent infrastructure in these areas, pervasive transmission networks can be installed. Besides the illumination, ventilation, air quality monitoring, and electrical energy supply networks already installed underground, we can design the communication network along these paths using fiber optic cable, coaxial cable, copper wire, or wireless mesh nodes.

Although there is a high potential to damage the network infrastructure, the placement and mounting of cabling and hardware along the permanent transport paths should be clearly considered due to the benefits of an integrated network of communication that brings higher security and safety of personnel underground along with higher profit for the mining companies.

The underground coal mines commonly have coal dust and methane gas present in suspension. This is the reason that all electrical and electronic devices (from illumination, ventilation equipment, transport means, and exploitation machinery) must be designed for intrinsic safety to minimize the risk of sparks. Sparks (e.g., that can rise from this improper-protected equipment) are the origin of methane gas or coal dust ignition in the mine, conducting to fire hazards with lethal accidents and important material and time loss.

Regardless of the local or remote communications solution, legislation must be adhered to, and only sanctioned "intrinsically safe" devices must be used underground. Prior to selecting a communication solution, the intrinsically safe network devices must be used in any mine (not only coal mines) that contains fire hazards.

Audio, video, and data communication systems are used underground in addition to or as a part of the dedicated network system. Gas and air quality monitoring and notification, geotechnical monitoring, blasting control, fire detection and alert, and emergency communication systems are installed underground according to the regulatory codes and vendor's requirements.

The existing communication systems underground should integrate a personnel positioning and monitoring system to enhance the mining company value. Both assets and personnel tracking underground require detailed planning to best achieve the objectives for a solid, reliable network.

Although such a system does not seem to have many benefits or high value in terms of productivity, it is worth to be implemented as a reliable live-saving system, for the security and personnel safety underground.

Each mine has its own topology and, therefore, requires a custom solution when we refer to a tracking system. Most of the positioning systems track the movement of personnel throughout spaces into the underground mine where the coverage infrastructure is mounted and notify on the personnel location. The tracking system should be tailored to cover all the spaces into the mine infrastructure underground [4].

Situational awareness, especially when supported by novel technologies (as virtual reality, augmented reality, or robots), is often associated with the tracking systems, and may assist personnel situated in the underground spaces to evaluate very quickly a potentially problematic situation and take in due time the best decision for her/his own safety and security [5–7].

In case of a duplex communication (e.g., tracking and warning system) the position of each miner in the underground spaces can be known by the integrated system and warn the individual or group of workers in due time about possible dangerous situations or areas under restrictions. Any integrated, personnel monitoring system digitally recorded, implemented in mines has many incontestable advantages over any classical manual record.

Such a real-time personnel monitoring/tracking system can instantly output a list with all workers and the identity of each worker situated underground, along with their locations. For instance, during an emergency, if an explosion with walls and roof collapse inside the mine takes place, the emergency response teams will focus

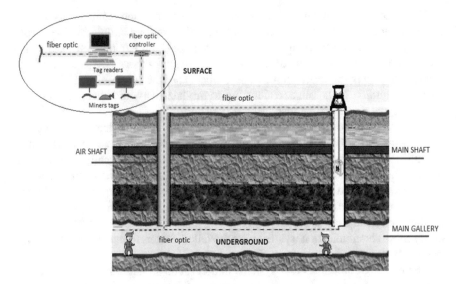

Fig. 3.2   A general view of a networking backbone of fiber optic

on affected areas where workers are located, considerably limiting the area of search and decreasing the rescue response times.

Unlike the open spaces outdoor, where public cellular networks or private satellite networks can be used for fast and accurate localization, in underground spaces, because of the lack of signal coming from satellites or cellular towers, most of the tracking systems developed so far, rely on low-frequency tags (BLE, ZigBee, 6LowPan, iBeacon, Eddystone, etc.) for real-time location, proximity location, or checkpoint positioning methods. These systems require a dedicated infrastructure in these cases and maps play an important role in any tracking strategy; therefore, the area of the site shown on a level map will need to be determined.

Most of the modern mines have an underground widely deployed communication infrastructure with optical fiber as networking backbone (Fig. 3.2), to support Wi-Fi or UWB technologies [8].

The coal mining industry has special conditions that impose very strict regulation regarding the use of any kind of electrical equipment. The lighting fixtures (as well as any electrical or electronic equipment), both on the main galleries underground and in all working spaces, are compulsory to be explosion proof, certified ATEX in order to meet the highest safety standards.

Explosion hazard involves the use of special lighting to avoid accidents because in coal mines, due to high firedamp or high level of methane concentration, atmosphere is favorable to explosion. Dusts, gases, vapors, and/or mists in air can form explosive atmospheres. Hazardous area classification is used to identify places where, because of the potential for an explosive atmosphere, special precautions over sources of ignition are needed to prevent fires and explosions.

Therefore, explosions can be caused by any mechanical or electrical spark. With the aim of preventing or minimizing the effect of explosions, beside mine ventilation, both electrical equipment used underground and illumination have to be specially designed to be safely used in the explosive atmosphere.

Any type of lighting and electrical equipment used underground must be able to withstand any possible internal explosion in order to prevent any risk of spreading an explosion, which is called explosion proof. So, workers' safety has to be in the first place when any type of VLC system prototyping is considered.

All types of lights (lighting fixture, miner's lamps, or machinery LEDs) used in coal mines have to comply with ATEX directive to avoid any risk of explosion or its propagation.

The ATEX normative for potential explosive atmosphere results according to European Directives: 1994/9/EC and ATEX 137 for any equipment intended for use in ATEX zones; 100A ATEX 1999/92/EC or safety of workers.

The most important condition when designing a VLC system is miners' safety, then, also essential, its efficiency, durability, low energy consumer, and long life since elements of both oTx and oRx have to proper operate in harsh environments of the mining industry [9].

Although the main purpose for this chapter is to describe the steps necessary for development of a local underground VLC wireless system, the remote communication—to the mine's surface, of the data acquired, has to be also considered since the continuous surveillance of the activity underground, and most important, the fast reply in case of emergency, is the final purpose of the entire communication network presented here.

3.2 System DesIGN

3.2.1 General Description

The designed system is an Underground Positioning & Monitoring System (UP&MS), based on the visible light wireless data communication, that can be used to individually identify each worker situated underground, as well as to monitor on a map, in real time, their position. The entire system refers to a real-time monitoring of the personnel located underground and also a surveillance mapping of the personnel situated onto the main galleries. The wireless optical medium of transmission relies on the visible light.

At present, in most of the traditional mines, the evidence of personnel to be found in underground mining spaces is a difficult one, based on hand-writing evidence, not that reliable. Data registered in this way can only attest personnel situated underground with certain identification but without any possibility to locate them. Data written on paper do not allow real-time monitoring and a certain identification of the position of personnel in horizontal underground mining spaces (galleries and rooms) vertical (mineshafts) and inclined (inclined planes, suitors, mineshafts), and so on.

Although today, identification of personnel situated underground is done due to a strict correlation between the protective equipment to be worn and personnel identification before they go underground, their localization in real time is not possible.

The specific conditions of mines that have a major risk of serious incidents, such as methane or pit-gas (fire-dump gas) explosions, shocks, fires, or accidents, endorsed rigorous internal rules. Underground access is only allowed with specific equipment authorized and approved by the competent mining national institutions.

The Mining Health and Safety (MHS) rules regarding the evidence of personnel situated underground states that, when entering underground, both staff and visitors are obligated to register themselves at the time of check-in at the "lamp office" of the mine where they receive the necessary equipment to be compulsory worn underground.

Here they will be registered, and if they remain underground from one shift to another, the shift supervisor will announce the workers who are underground to be registered in the Underground Personal Records Register, then forward the same information to the MHR responsible employee. This scripted monitoring and highlighting system is a difficult one that does not provide a real-time reporting of the number and position of people underground at a given time.

The number of people situated underground is in direct correlation with the authorized equipment mandatory to be worn underground, that staff receive before they go in mine, namely clothing, footwear, personal protective equipment, and individual lighting equipment.

The protection equipment considered for the system implementation consists of the miner's helmet cap-lamp. This has a unique identification number which is correlated, each time, with the person that is wearing it in the underground mine. The designed system introduces a simple identification of the miner's helmet cap-lamp based on the EAN-8 code bars, and therefore of the person that uses that specific lamp, underground.

A general view of the entire positioning and monitoring system with the VLC local wireless communication and remote transmission to the surveillance room is presented in Fig. 3.3.

The system has as the main goal, to identify in real-time, of each person in the underground mine spaces, based on the miner's helmet cap-lamp as the main "actor" of the system.

Since both lamps and helmets are compulsory to be used underground, they are nonstop worn by personnel during movements from one place to another. Communication of the position of worker's underground is done by a transmitting system that uses, in its first stage of data acquisition, the technology of the visible light, as local wireless data communication means.

The oTx device is embedded into the miners' helmet cap-lamp (Fig. 3.4); the lighting device of the lamp has to be most of the time positioned on the worker's helmet. The lighting device of the lamp can be easily tilted and therefore changed, to fit a precise light beam needed at any time. Moreover, the lamp has lenses embedded that can focus and direct the light as the worker wishes.

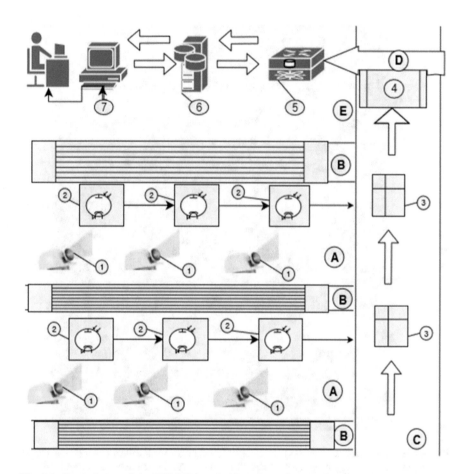

Fig. 3.3 General view of the UP&MS installed into the underground spaces. (1) Miner's helmet cap-lamp as oTx with the VLC driver system embedded. (2) Access Point as oRx with the VLC driver embedded. (3) VLC location controllers. (4) Master controller board—Communication head node. (5) Router. (6) Server. (7) PC. (a) Main galleries with the illumination network/optical fiber already setup. (b) Coal/waste seam. (c) Vertical mine shaft. (d) Operating cage room. (e) Surface main monitoring and surveilance room

The beam of light from the miners' helmet cap-lamp sends, piggybacked by illumination, the information (using IM/DD method) related to the lamp's ID, namely its own unique EAN-8 code bar. This code bar—converted in bits and sent by beam light due to the light shifting from ON (bit 1) to OFF (bit 0) and vice versa at high speed—is received by the Access Points (APs aka oRxs). Each AP consists of an electronic driver device with a microcontroller and a PIN PD with optics attached, all of them embedded into the lighting network on the main galleries underground or into the optical fiber network backbone.

Data regarding the lamp ID are piggybacked by light to the APs (with the VLC embedded) that act as visible light communication receivers (oRx) (Fig. 3.5).

Fig. 3.4 Both VLC transmitter (oTx) and receiver (oRx) on the main gallery underground. oTx—on the miner's helmet cap-lamp. oRx—embedded into the illumination network/optical fibre

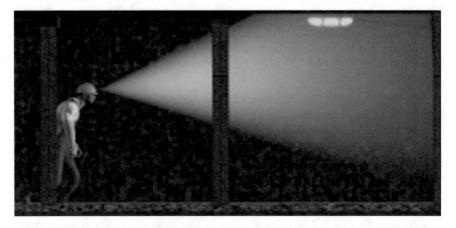

Fig. 3.5 The light beam of the miner's helmet cap-lamp, piggybacks data to AP

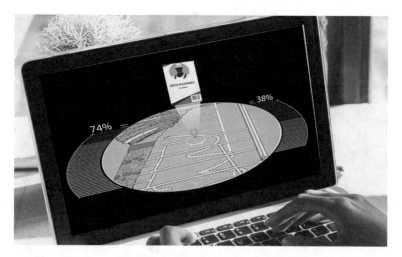

Fig. 3.6 The miners' identity and her/his location displayed on a digital map

The oRxs can be incorporated either into the illumination network already set up on the ceiling of the main galleries in the underground spaces, or connected by the optical fiber network, that acts as the backbone for multiple remote communication systems.

The lamps' ID/code received here is automatically checked having the last digit that as checksum (CS). CS, as an error-detecting element, will check the code to detect the correct received data.

During the next stage, each AP adds its own identification data (its own position on the gallery, embedded in the system, the date and time of received information from the oTx) into an Ethernet frame type II that is going to be sent (through the lighting network or the optical fiber network), to the surface, at a dedicated PC situated into the main monitoring room of the mining company.

Data are collected in the location controllers (positioned at the same, optimum distance), and are forwarded to the master controller board which is the communication head node.

All these devices can be situated on the illumination network or on to optical fiber networks, which act as the backbone for multiple remote communication systems.

Data acquired in the main galleries and working spaces, by the oRx are sent from the APs through the illumination system/optical fiber to the main shaft (C) (see Fig. 3.3) where the controllers are situated and then to the surface, at the operating cage room. From the operating cage room, it is networked to the main monitoring and surveillance room. Data are finally collected and stored into the mining company server, and the personnel position in the underground can be seen by the supervisor, in real time, both as records lists and on a digital map (Fig. 3.6).

This system is useful not only because the position of each worker situated in the underground is known, being recorded and displayed on a wide screen in the surveillance room, but the identity of each worker is also known.

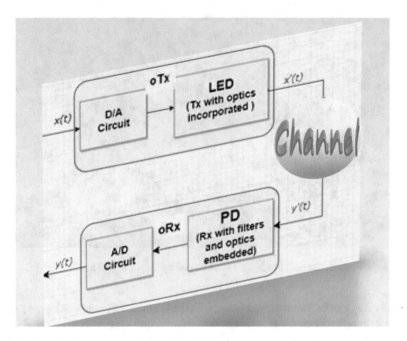

Fig. 3.7 The functional representation of the VLC system

The UP&MS system described here is a hybrid communication network, consisting of a local wireless communication, where data are sent through light and a remote cabled transmission network that sends data in real time to the surveillance room. Most of the emergency communication networks are designed as to resist into the hazardous mine environment since the impact of mine hazards such as rock falls, floods, fires, and explosions can cause catastrophic failures of the system underground. An emergency system is required to maintain service in spite of component failures, which is unlikely in most of the real situations. The system we propose allows a continuous surveillance (both in real time and digitally stored) of the workers underground and keeps valuable information about their identity and position in underground spaces prior to any possible disaster to occur, making therefore the rescuers' mission more accurate.

The Ethernet technology as cabled (cooper or optical fibre) network has already been deployed in major mining companies worldwide, therefore, the system description is going to be focused on the local wireless data transmission through visible light.

A functional representation of the local wireless VLC is presented in Fig. 3.7.

Data regarding the lamp's ID ($x(t)$) are able to be sent due to a digital to analog (D/A) signal converter circuit and the LED with the default optics embedded resulting a signal ($x'(t)$), ready to be send through free space to the oRx.

The channel allows the optical signal to be sent to the oRx. The oRx module, with a PIN PD and optics (filter and lens) embedded, as the receiver, converts the signal from analog to digital (A/D) form, due to the circuit that drives and converts data received ($y'(t)$) from the oTx.

The electronic circuit with microcontroller checks the digital information received (the lamp's ID) and then builds the Ethernet frame (adding few more information, as its own AP's position, time, and date of data received) to be sent to the main surveillance room in the surface area, on the path settled as an underground and surface local area network LAN.

The most important function of the miner's cap-lamp is illumination, as underground lighting is a very important, fundamental factor. Therefore, designing the VLC system for personnel positioning and real-time monitoring has to take primary in consideration illumination as a key aspect, since in underground mines, it is vital to have safe and efficient lighting.

The IM/DD technique is the suitable method to communicate optical data underground in a VLC system, due to its low cost and simplicity. The proposed system implements the EAN-8 code bar as it has a low cost, and is a simple one, being therefore, a proper choice for a viable underground positioning and monitoring system.

The EAN-8 code bar is visible printed on a label as a sticker on the lamp and protected with a transparent layer to be easily scanned. The same code it is also digitally stored into the VLC oTx driver. A dedicated database stores on the server (no 6—see Fig. 3.3) all EAN-8 codes available for all lamps in the company, therefore each lamp has its own ID.

When the personnel (staff, workers, visitors, etc.) receive the lamp, his/her identity is directly linked to the lamp's ID from this moment till he/she returns the lamp at the mine's surface. The database will update this link and therefore personnel underground will be not only monitored, but all the time identified, and their position in a certain moment known, as well.

The LED on the cap-lamp is ON all the time, both on the way to and from destination when employees, staff and/or visitors walk underground on the main galleries to the working spaces. During all this time, the oTx with VLC driver embedded will continuously send its own data consisting in the identification number (ID), aka its EAN-8.

The system is considered to be the best possible choice at this moment in order to be reliable in case of accidents underground, being extremely useful to identify the actual or latest location where the personnel have been right before or during the accidents.

The EAN-8 code consists of seven usable decimal digits and a check digit as presented in Table 3.2.

An example of EAN-8 code is presented in Fig. 3.8.

The first digit will identify the level of responsibility in company, according to each employee function or profession (staff, workers) as well as visitors.

There are ten possible levels of classification (digits from 0 to 9) according to the internal rules of each company or any specific identification requirements, therefore, the first decimal will be assigned as default ID.

Table 3.2 EAN-8

Character set	Length	Check digit	Size, Module Width X, Print ratio
numeric [0..9]	7 usable digits	1 check digit	Font size SC2 (SC0–SC9); $H = 21.64$ mm (17.7–43.28); $B = 26.73$ mm (21.87–53.46); $X = 0.33$ mm (0.27–0.66); variable size between 80% and 200%
Notes	EAN8 data consist of 2–3 digits of country code and 4–5 digits of article code (limited numbers)		

Fig. 3.8 EAN 8 bar code, an example

2345 6785

Table 3.3 Electrical and photometric characteristics of lamps used by miners

Miner's helmet cap lamp	Electrical Characteristics Supply Voltage (V$_{DC}$)	Current (A)	Power (W)	Photometric Wavelength (nm)	Correlated color temperature (CCT) (K)
Lamps with incandescent bulbs	6.1	0.63	3.84	780	2880
Regular LEDs	6.1	0.42	2.56	448	5855
Prototype LED	12.0	0.113	1.36	444	6844

The other six digits (any decimal from 0 to 9) consist of the lamp's ID. The last digit is the checksum (CS) digit.

LEDs are the most used artificial lighting means recently due to technical advantages of this technology such as efficiency, increased longevity, and energy savings.

The designed LED cap-lamp prototype must also allow, an improved visual performance since the miners are more sensitive to glare. According to the NIOSH and MSHA, new LED lamp prototypes recently used in mines are robust, safe (they do not have a filament that can break or a glass envelope), have a long life (50,000 hours of operation) and "improve the ability of older miners to detect moving hazards, by 15 percent, and trip hazards by nearly 24 percent" [10].

According to Sammarco et al. [11], LEDs with shorter wavelengths are able to offer improved peripheral motion detection, reduce glare for elder workers, and avoid floor hazard early detection.

Table 3.3 presents electrical and photometric characteristics of few types of lamps used by miners underground for the study, according to Sammarco et al.:

Fig. 3.9 Type of miner's cap lamp and its position on miner's helmet in the VLC system design considered

- Lamps with Incandescent bulbs.
- Lamps with Regular LEDs.
- Prototype LED.

Following all considerations above, modeling and simulation results obtained and presented in the previous chapter, regarding the general and particular requirements for a proper development of the underground positioning and monitoring system based on VLC, both hardware and software optimum choice for the special condition in a potential explosives atmosphere and dense polluted medium as the underground coal mine environment is, are presented here.

The type of miner's cap lamp with LEDs (and portable battery attached), intended to have the oTx embedded, is presented in Fig. 3.9.

3.2.2 System Simulation

The Optical System

LEDs are recently the most used artificial lighting in the modern underground mines (for the miners' cap-lamp, as lighting fixtures into the illumination infrastructure or for machinery) due to their technical and economic advantages, as high efficiency, increased longevity, low cost, and energy savings.

There are two possible approaches regarding the type of diode used for such a VLC system. Either LEDs or SLDs can be used in both semiconductors.

LEDs and SLDs are based on the SSL technology that converts the electrical signals into light waves. Starting with the basics, diodes are the simplest form of a semiconductor.

The light beam produced by a LED has a wider FoV than the one produced by an SLD which has its light beam directed and highly collimated. Therefore, because the optical signal sent by LED is broader and is spreading, its optical output power is low and travels a shorter distance than the light beam produced by SLD which has a higher optical power.

In case that the infrastructure has narrow access paths in spaces underground, where only one worker can walk to the path, the SLD is a good choice and the system with SLD in an LoS setup can be a good solution, especially for the advanced optics embedded that minimize the blindness of workers because of the focused and high intensity of light beam emitted by SLD. On the other hand, when the access path is wider, and the worker is not forced by the type of infrastructure to walk under the backbone of the network where access points are situated, LEDs with a wider light beam should be used. On the oTx side, the collimating lens will reduce the beam's divergence to a level that allows the optical beam to maintain its integrity as it reaches the active area of the PD of the oRx.

When considering a VLC setup, there are several physical limiting factors that have to be overcome to improve the overall effectiveness of the system. The key idea is to collect a valuable incident light beam on the oRx and provide a more concentrated spot on the active area of the PD.

When the system of optics in front of the oTx is designed, it is very important not to forget that the main functionality of the miners' helmet cap-lamp still has to be lightening his/her own walking path with a proper light beam that has to comply with all the safety standards. Therefore, bearing in mind this, we also have to choose the diode and design the optics as not to raise the oTx system costs, weight, volume or light efficiency, and keep the system energy still efficient.

To comply with the LEDs and SLDs' safety standards, the entire oTx module used in a VLC system for wireless data transmission has to be designed accordingly. From the point of view of the system, solely, it would be easy to use an SLD that has a high output, which would take a lot of strain off the PD, and would increase the working distance considerably. However, to comply with the safety regulations, SLDs have to be relatively low in power, and then, to improve the distance and keep the optical power high, focusing on the elements that can be used as compensation.

When the system is designed as a duplex communication system, the two photodiodes on either end serve as the receiver for their respective devices assisting them in the collection of photons that improves their efficiency.

The LD beam can only have typical irradiance distributions due to the restriction of laser cavity and gain distribution. However, in practical application, the LD beams with desired irradiance distributions are often required to improve the efficiency. The basic principle of the aspheric lens (Fig. 3.10), in a refractive beam shaping system, is to allow the first aspheric lens (the one in front of LED) to redistribute the irradiance of the input beam to the desired distribution at the second aspheric lens plane (in front of PD) that re-collimates the output beam [12].

In Fig. 3.10, the LED's light beam diverges by an angle θ over the length d. A lens is placed in front of the LED and another one is placed at the length d, in front of the PD to intercept the beam and focus it onto the active area of the PD.

For the oTx side, in front of the LED, a simple aspheric plano-convex lens collimates the light beam. An aspheric lens, as the name suggests, is shaped in such a way that the surface is not one single radius, nor is it continuous.

Prior to selecting the proper lenses for an efficient light beam in a VLC design, we must investigate the off-the-shelf, available low cost refracting lens and systems.

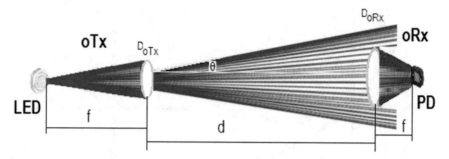

Fig. 3.10 A VLC system with lens in front of LED and PD

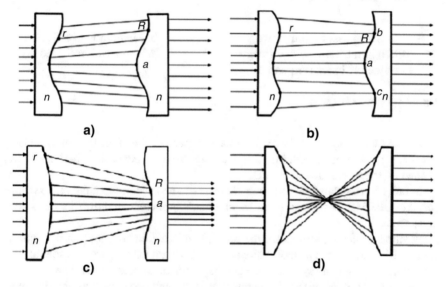

Fig. 3.11 Refractive light beam shaping systems with aspheric lens. (**a**) Configuration of a Galilean magnifying beam shaping system. (**b**) Configuration of Galilean demagnifying beam shaping system consisted of one magnifying and one demagnifying system. (**c**) Configuration of a Galilean demagnifying beam shaping system consisted of only one demagnifying system. (**d**) Configuration of a Keplerian plano-convex lens as refractive beam shaping system

Lenses used must improve both the light beam efficiency and the distance between oTx and oRx on its traveled path. Both the light beam ray tracing and its efficiency on a long distance have to be well established with the help of correct selected types of refractive shaping lens (or system of lenses) at the oTx and oRx sides.

According to the working principle, the refractive shaping system can be divided into two categories: Galilean (Figure 3.11a, b, c) and Keplerian (Figure 3.11d) light beam shaping systems [13].

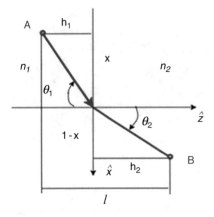

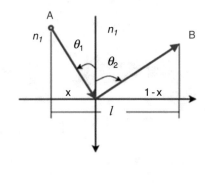

a. Ray passes through two different mediums with different refractive indexes n_1 and n_2

b. Medium with refractive index n_1 with reflected ray

Fig. 3.12 Fermat's law setup

All the shaping systems above give us a clear perspective of the kind of the lenses necessary to be used in order to obtain a long distance, beam light that has to be spotted on the active area of the PD.

In order to calculate the most suitable lenses to be used in the final VLC system for the UP&MS described here, the Fermat's principle and Snell's laws are considered.

Fermat's principle, also called the principle of least time, states that the optical rays of light traverse the path of stationary optical length with respect to variations of the path, meaning that rays take the path that requires the least travel time.

Figure 3.12 shows the case when raw travels through two mediums with different indices of refraction, n_1 and n_2. This is the case (Fig. 3.12a) when ray goes out of the lens (lens is of plastic with n_1 of about 1.6) and enters into the air.

Clean air has the refraction index $n_2 = 1$ but the underground environment, the authors consider in this work, is filled with tiny suspended particles of coal and rock within the air that continuously moves due to the air cleaning procedures that have to be strictly applied underground to follow rules for human safe and secure work.

The refractive index for coal is 1.85 (for anthracite) [14]. Different measurements, more accurate, show that the index of coal spreads within the range 1.68–2.02 (58–96% carbon coal) [15].

$$S_{AB(a)} = n_1 \sqrt{x^2 + h_1^2} + n_2 \sqrt{(1 - x)^2 + h_2^2} \qquad (3.1)$$

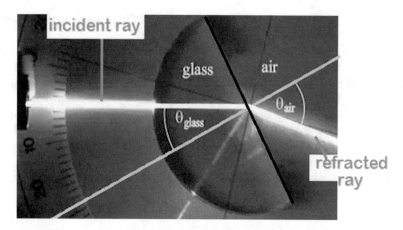

Fig. 3.13 Incident and refracted ray through glass

The light ray in the Fig. 3.12(b) starts from point A and reflects off the surface before arriving at point B. Let's consider l as the horizontal distance from point A to B.

In order to find out the time required for the light to travel between the two points, the length of each path is calculated and divide the length is divided the speed of light (Eq. 3.5). Therefore, refraction is due to the speed of light through two different mediums (Fig. 3.13).

$$t = \frac{n_1 \sqrt{x^2 + h_1^2}}{c} + \frac{n_2 \sqrt{(1-x)^2 + h_2^2}}{c} \tag{3.2}$$

When air is highly polluted nearby the working spaces, the atmosphere becomes almost opaque and therefore the light beam rays hit the tiny particle of coal in the air and reflect from their surface. The most important properties of light in this type of polluted medium are absorption and scattering.

Attenuation (or transmission loss) refers to the intensity of the light beam that decreases with respect to distance traveled through a communication medium, therefore an extinction coefficient has to be taken into consideration when the polluted environment is measured.

Attenuation (A) coefficients use units of dB/m through the medium:

$$A = 10 * log_{10} \frac{I_{input}}{I_{output}} \quad (dB) \tag{3.3}$$

Optical attenuation is caused both by scattering and absorption. Attenuation is an important factor limiting the transmission of an optical signal across long distances. Thus, worldwide research efforts are concentrated on both increasing the efficiency of the emitted light beam and limiting the attenuation of the optical signal on its path between the oTx and oRx.

The medium considered here starts from a clean air in most of the parts of the main gallery (due to the air cleaner procedures imposed underground with dedicated equipment) to highly dense air polluted with fine particles of coal and rock closer to the working space.

According to Snell's law of refraction, when a light ray passes from one medium (e.g., glass) to another one (e.g., air), as the angle of incidence increases, the reflected ray becomes stronger and the refracted ray becomes weaker and eventually disappears.

The relation between the incident and refracted ray through glass is as in Eq. 3.4.

$$\frac{\sin \theta_{air}}{\sin \theta_{glass}} = \frac{n_{glass}}{n_{air}} \tag{3.4}$$

where:

n_{glass}—refractive index of glass

n_{air}—refractive index of air

Biconvex lenses (Fig. 3.14) have a positive focal length and converge the incident light. They have symmetrical form, with equal radii on both sides. These are used for the purpose of virtual imaging in case of real objects and also for a positive conjugate ratios ranging between 0.2 and 5.

The optical path length S, according to Fermat's law, is presented in Eq. 3.5:

$$S = S_1 + S_2 = -\frac{r_2^2}{2f} - \frac{r_1^2}{2f'} = -\frac{r^2}{2}\left(\frac{1}{f} + \frac{1}{f'}\right) \tag{3.5}$$

$$\frac{1}{F} = \frac{1}{f} + \frac{1}{f'} \tag{3.6}$$

Power of lenses adds as:

$$\varnothing = \varnothing_1 + \varnothing_2 - \frac{d}{n}\varnothing_1\varnothing_2 = \frac{1}{r_1}(n-1) + \frac{1}{r_2}(1-n) + \frac{d}{n}\frac{1}{r_1 \cdot r_2}(n-1)^2 =$$

$$= c_1(n-1) + c_2(1-n) + \frac{d}{n}c_1c_2(n-1)^2 \tag{3.7}$$

Three scenarios of simulation done with support of geometric-optics module on *Phet* [16] are shown in Figs. 3.15–3.18.

As general conclusions, the shape of a lens, its curvature radius, the material of its composition (glass, or plastic), the distance between LED and lens, the focal length, the diameter of the lens as well as the alignment between LED and lens are just few of the key characteristics that have to be considered and fine-tuned to obtain the optimum optical setup for a reliable wireless VLC system.

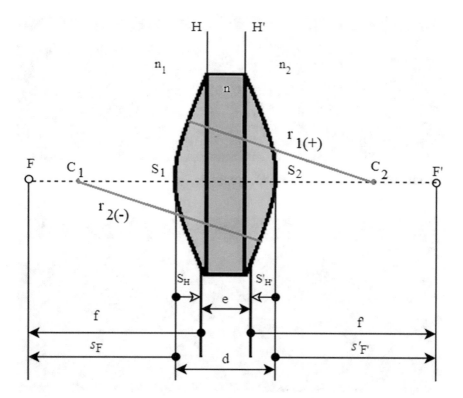

Fig. 3.14 Biconvex lens

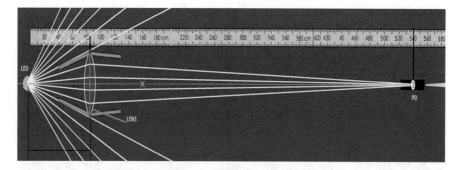

Fig. 3.15 Biconvex lenses in front of LED, case A

1. When the LED and LENS are perfectly aligned, for LENS with a curvature radius of 0.3 m, a refractive index of 1.2 and diameter of the biconvex LENS of 0.58 m, rays converge at a distance (away from the LED) of about 540 cm on the PD's surface. The distance between the LED and lens is about 85 cm (Fig. 3.15).
2. With the same scenario, when the LED and LENS are perfectly aligned, with the same LENS curvature radius of 0.3 m and refractive index of 1.2 with diameter of

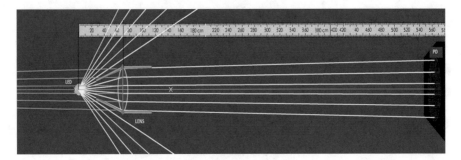

Fig. 3.16 Biconvex lenses in front of LED, case B

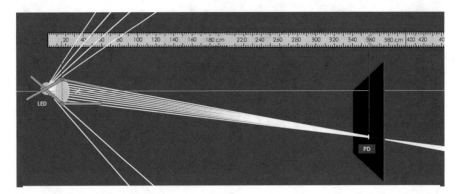

Fig. 3.17 Biconvex lenses in front of LED, case C

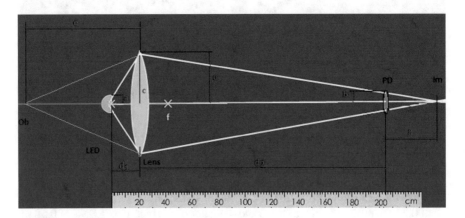

Fig. 3.18 Ray tracing according to Phet simulation

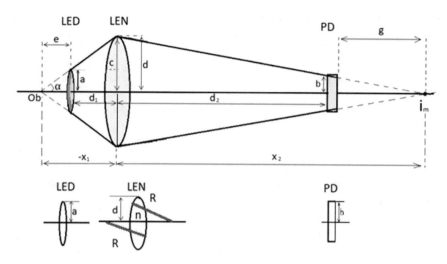

Fig. 3.19 Schematic representation of LED, biconvex lens (LEN) and PD

the biconvex LENS of 0.58 m, moving the biconvex LENS, to the left, closer to the LED, rays become all parallel (Fig. 3.16) and at the backside to the LED, image is forming (green rays).

On the other hand, when the refractive index becomes 1.84, with the same curvature radius of 0.3 m and the diameter of the biconvex LENS 0.3 m, with the distance between LED and the biconvex LENS of 0.18 m, rays converge on the PD surface at a distance of 358 cm (Fig. 3.17). This time LED and LENS are not more perfectly aligned. The image is formed on the PD only when it is placed in the direction of the rays (with an elevation angle $>0°$) formed in this way.

Simulations in Figs. 3.15–3.17 give us insight about the distribution and path of the rays of light in different scenarios with a biconvex LENS in front of the LED.

In fact, the LED's aria is much smaller; therefore, the lens diameter has to be accordingly smaller. Using mathematical models and simulation, taking into account values closer to the reality, optimum position of LENS related to the LED can be found.

In Fig. 3.18, a curvature radius 1.37 m, lens refractive index 1.87, diameter of biconvex LENS 0.76 m and 3.19, a LED, biconvex LENS, and a PD are considered.

Based both on the mathematical model and Fermat's principle and Snell's law, according to the following Fig. 3.19 and demonstration, as well as simulation, the optimum distance between LED and lens (LEN) can be established with a high level of accuracy.

a—LED's semidiameter,
b—PD's semidiameter,
d—semidiameter (marginal ray) of lens,
f—focal distance of lens,

R—radius of curvature (biconvex lens).
d_1—distance from LED to lens center (LEN),
d_2—distance lens to PD,
c—semi-length of ray in lens,
e—distance from Ob to LED,
g—distance from PD to I_m.

Next, will be considered:

$$c \le d \tag{3.8}$$

Since any of the light rays should not be lost, most, or even all the optical power from LED has to be used.

$$\frac{1}{f} = (n-1)\left(\frac{1}{R_2} - \frac{1}{R_1}\right) = (n-1)\frac{2}{R} \tag{3.9}$$

$$f = \frac{R}{(n-1)2} \tag{3.10}$$

x_1—the left side from the LED, according to convention, has negative value <0

$$-x_1 = e + d_1 \tag{3.11}$$

Taking into account the similarity of the triangles in Fig. 3.18 results Eq. 3.12.

$$\frac{e}{d_1} = \frac{a}{c-a} \tag{3.12}$$

For:

$$tg\alpha = \frac{a}{e} \quad => \quad e = \frac{a}{tg\alpha} \tag{3.13}$$

In case that $c = d$

$$\frac{e}{d_1} = \frac{a}{d-a} => d_1 = \frac{e \cdot (d-a)}{a} = \frac{a}{tg\alpha} \cdot \frac{d-a}{a} = \frac{d-a}{tg\alpha} \tag{3.14}$$

From Eq. (3.14)

$$=> -x_1 = e + d_1 = \frac{a}{tg\alpha} + \frac{d-a}{tg\alpha} = \frac{d}{tg\alpha} \tag{3.15}$$

As it can be seen in Fig. 3.20, $x_1 > f$.
Taking into account the similarity of the triangles in Fig. 3.21 results Eq. 3.16.

Fig. 3.20 The two triangles alike behind lens

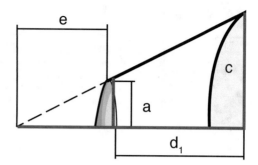

Fig. 3.21 The two triangle alike in front of lens

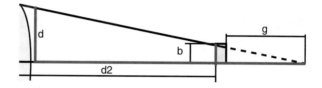

$$\frac{d-b}{b} = \frac{d_2}{g} \tag{3.16}$$

$$g = \frac{d_2 b}{g} \tag{3.17}$$

$$x_2 = d_2 + g = d_2 + \frac{d_2 b}{d-b} = d_2\left(1 + \frac{b}{d-b}\right) \tag{3.18}$$

$$x_2 = d_2 \frac{d}{d-b} \tag{3.19}$$

According to Eq. 3.18

$$\frac{1}{x_2} - \frac{1}{x_1} = \frac{1}{f} \tag{3.20}$$

$$\frac{d-b}{d_2 d} + \frac{tg\alpha}{d} = \frac{1}{f} \tag{3.21}$$

$$d_2 = \frac{f(d-b)}{d - f tan\alpha} \tag{3.22}$$

Figure 3.22 shows a simulation made with the aim to find the optimum distances d_1 and d_2 according to LEDs, LENS', and PD's characteristics, positions of LED, LENS, and PD related to each other, and Snell's law.

Taking into consideration convention, results $d_1 = -99$ and $d_2 = 128.6$. The results obtained here based on the mathematical models and simulation constructed with the support of Simulink are similar with the results simulated with the support

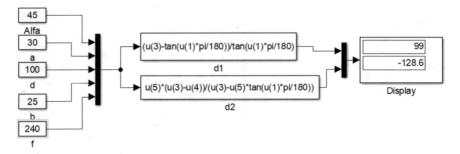

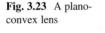

Fig. 3.22 Simulation for distances d1 and d2 in case of a biconvex lens placed in front of LED

Fig. 3.23 A plano-
convex lens

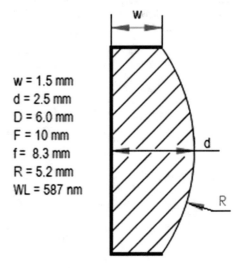

w = 1.5 mm
d = 2.5 mm
D = 6.0 mm
F = 10 mm
f = 8.3 mm
R = 5.2 mm
WL = 587 nm

of Phet app; therefore, the geometry, dimension, position, and type of lens in front of LED can be clearly determined for a clean air optical channel.

Depending on the system designed, a proper choice for focusing the light beam that hits the PD into the oRx is a short focal length plano-convex lens (Fig. 3.23) which is designed for the peak laser wavelength. The reason is that a well-collimated beam can be used and neatly readjust the beam into a smaller spot size, reducing the stray light and improving optical radiation incident onto the oRx.

The plano-convex lens in Fig. 3.22 is designed for 587 nm wavelength with diameter $D = 6$ mm, the focal length $F = 10$ mm, the curvature radius $R = 5.2$ mm, and the back focal length is $f = 8.3$ mm.

Prior to purchasing and testing distinct types of lenses with various characteristics, a very useful tool is the OpticStudio [17] application where the proper optical system can be simulated. This is one of the most advanced, dedicated optical design software which allows the user to make an infinite sum of an optical train lenses and see the results. For example, the OpticStudio Sequential Mode allows a basic process of designing a lens, building a dedicated system, analyzing its performance, and optimizing it for the required prescription and design constraints.

There are a series of specifications (focal length, semi FoV, wavelength, center thickness of the lens, edge thickness of the lens, and object location) and constraints of the application when designing and optimizing a single lens made of glass.

The simulator is a calculator that performs the commands given, however, intrinsic in the design of the program is an optimizer which can pick the best criteria of selection such as spherical, coma, or astigmatic aberrations. The RMS spot size is a metric for visually understanding the effects of stray light or aberrations on detection quality. The smaller the spot size, the more control of the light to focus.

Important features of the OpticStudio packages are the CAD integration, programming interface, lighting and illumination design, or data libraries (lenses, materials, coatings, radiant sources, scatter, or spectrum data files).

The Electrical Drivers

The oTx Electrical Circuit and Simulation with Microcontroller

The first step to be taken into consideration when designing a circuit with LEDs is that too much current and voltage may damage the LED. The simplest way to protect a LED is to include a current limiting resistor in series. LEDs have recommended operating voltage and current, based on which resistance can be calculated using both Ohm's and Kirchhoff's laws.

A very useful and powerful tool, the online *EveryCircuit* [18] application offers visual support in order to see, in real time, the LED's voltage and current displayed on the oscilloscope, therefore data can be easily plotted and optimum values identified.

The current can be seen as represented in the circuit simulation with green dots, since it is a flow of electric charge, such as electrons moving through a wire. Its amount measured in amperes (A), as well as its direction (defined to be the direction of the flow of positive charges) is represented on the circuit. A current source of five amperes, for example, transports five coulombs of charge per second. The online application *EveryCircuit* displays, by default, the conventional current and the moving green dots representing positive charges. Since electrons carry negative charge, they move in the opposite direction. On the application, the speed and brightness of the dots represent the amount of current.

The circuits are built using 2 N3904 NPN Silicon transistors and 2SC5200 NPN Silicon transistors.

The important characteristics of the 2 N3904 NPN Silicon transistor that have to be taken into consideration for system design are $V_{CC} = 3.0$ V_{dc}, $V_{BE} = 0.5$ V_{dc}, $I_C = 10$ mA$_{dc}$, $I_{B1} = 1.0$ mA$_{dc}$, $t_{he} - 35$ ns).

Its rise time ($t_r = 35$ ns), the storage time ($t_s = 200$ ns), and its fall time ($t_f = 50$ ns) are also important features to consider.

The diagram and the equivalent test circuits are presented in Fig. 3.24.

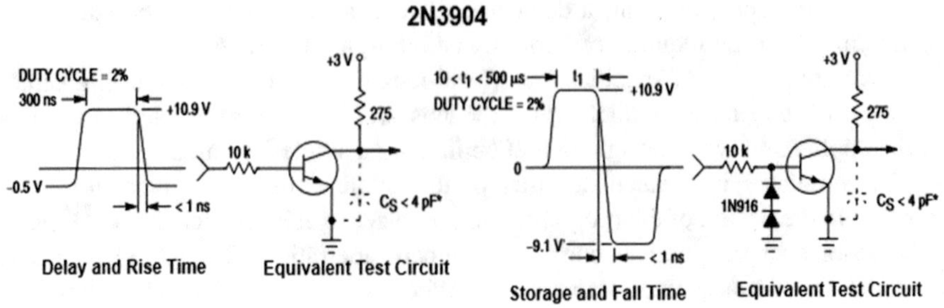

2N3904

Delay and Rise Time   Equivalent Test Circuit

Storage and Fall Time   Equivalent Test Circuit

Fig. 3.24 Diagram and equivalent test circuits for 2 N3094 transistor

Table 3.4 Characteristics of NPN transistor 2N3904

DC Current Gain		Min	Max
$I_C = 0.1$ mA$_{dc}$, $V_{CE} = 1.0$ V$_{dc}$	h_{FE}	40	–
$I_C = 1.0$ mA$_{dc}$, $V_{CE} = 1.0$ V$_{dc}$		70	–
$I_C = 10$ mA$_{dc}$, $V_{CE} = 1.0$ V$_{dc}$		100	300
$I_C = 50$ mA$_{dc}$, $V_{CE} = 1.0$ V$_{dc}$		60	–
$I_C = 100$ mA$_{dc}$, $V_{CE} = 1.0$ V$_{dc}$		30	–
Collector–emitter Saturation voltage			
$I_C = 10$ mA$_{dc}$, $I_B = 1.0$ mA$_{dc}$	V_{CE}	–	0.2
$I_C = 50$ mA$_{dc}$, $I_B = 5.0$ mA$_{dc}$		–	0.3
Base–emitter Saturation voltage			
$I_C = 10$ mA$_{dc}$, $I_B = 1.0$ mA$_{dc}$	V_{BE}	0.65	0.85
$I_C = 50$ mA$_{dc}$, $I_B = 5.0$ mA$_{dc}$		–	0.95

Main features, the electrical characteristics, of Q1 2N3904 NPN Silicon transistor (first on the left in Fig. 3.24) that are important for simulation's result, are presented in Table 3.4:

The 2SC5200 NPN epitaxial silicon transistor is a general-purpose amplifier, manufactured by Toshiba company, with high transition frequency of $f_T = 30$ MHz, large current $I_C = 15$ A, being ideal for use as complementary transistors in a Darlington pair configuration for amplifier applications also in applications as amplifier audio output high fidelity powered 100 W RMS.

The difference between collector and base or emitter can reach up to 230 V, between base with emitter 5 V, power gain (h_{FE}) between 55 and 160 at $V_{CE} = \pm 5$ V and $I_C = \pm 1$ A.

The main electrical characteristics of SC5200N BJT transistor are presented in Table 3.5:

The electronic circuit uses a Darlington transistor configuration of two NPN bipolar transistors 2 N3904 and 2SC5200 in order to increase current switching for a given base current.

Table 3.5 Characteristics of NPN transistor SC5200N

Characteristics	Symbol	Unit	Max
Collector–emitter breakdown voltage	I_{CBO}	μA	230
Collector–emitter saturation voltage	V_{CE} (sat)	V	3
Base emitter voltage	V_{BE}	V	1.5
Transition frequency	f_T	MHz	30
Collector output capacitance	Cob	pF	200

The first scenario simulates the use of three cold white LEDs (1 W each) with a 10 KΩ potentiometer in a circuit, in order to determine the LEDs' highest brightness and lowest dimming values levels as seen in Figs. 3.25 and 3.26.

Simulation of the electronic circuit of the oTx in this first scenario (with three cold LEDs of 1 W) is presented in Fig. 3.25.

This first simulation has the main aim to establish the LEDs' maximum brightness as well as their dimming, according to the maximum current possible on each LED (350 mA). All these data are important to identify the cut-off frequency in the developed VLC system.

On the application screen, all the passive and active elements (potentiometers, resistances, transistors LEDs, etc.) can be tuned until LED's both voltage and current are ideal. On the screen, LED lights up till it has enough current and also indicates when the current exceeds the nominal by over 2× (Fig. 3.25).

The 10 KΩ potentiometer used into the circuits (Figs. 3.25 and 3.26) is used to control LEDs' brightness/dimming level.

Here, the variable resistor (also called potentiometer or pot) has the effect of reducing/increasing the optical signal; therefore, it acts like a signal's attenuator/amplifier.

Beside values already seen on the circuit, data are plotted on each part, as follows:

LED—cold white 1 W (maximum 3.3 V and 350 mA).
R1—potentiometer of 10 KΩ (resistance plotted of 100 Ω).
Q1—NPN transistor with forward beta 32.2 with a 29.6 mA.
Q2—NPN transistor with forward beta 60 with a 1.8 A.

LED 1, 2, and 3—601 mA is higher than 350 mA, the maximum tolerated by our type of LED

When testing the highest brightness of the three LEDs, at a 12 V supply, the values above elements are higher than the limit supported by the three LEDs and thus LEDs are completely damaged, as it can be seen on Fig. 3.25.

At 12 V charge the current is 919 mA for an NPN transistor (Q1) with forward beta of 12.6 (15.1 mA) and an NPN transistor (Q2) of forward beta of 60 (919 mA), the current on each of the three LEDs is 306 mA at 2 V.

The circuits above, however, do not use the OOK modulation technique in order to send wireless data piggybacked by light.

A different design enforces LEDs to flicker at the highest frequency allowed by their technical characteristics established by the manufacturer.

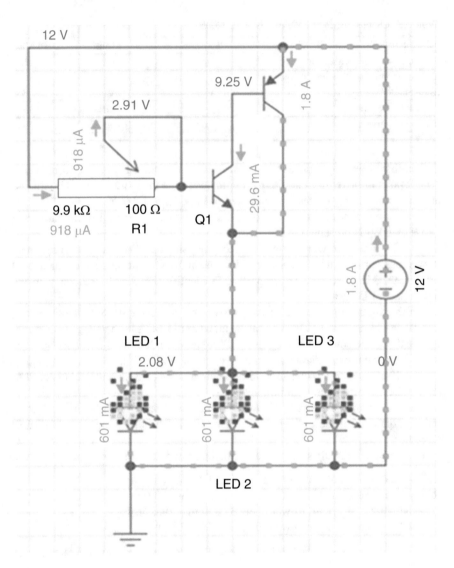

Fig. 3.25 Simulation of circuit with three LEDs dimming/brightness with overloaded current on LEDs

Therefore, Fig. 3.26 presents the proper design of the circuit for the purpose mentioned above.

Designing a proper circuit enforces, first of all, to meet the constraints imposed by the environment where the system will work.

Since most of the miner's cap lamp models tested (with the oTx embedded) are charged at 5 or 3.3 V, the circuit has to be modified accordingly. On the other hand, there is one cold white LED (1 W) with illumination purpose (positioned into the

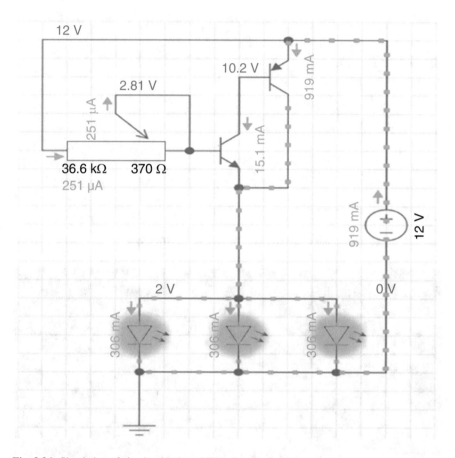

Fig. 3.26 Simulation of circuit with three LEDs dimming/brightness

center of the illumination box) and four other different LEDs with different other purposes.

Besides the elements used in Figs. 3.25 and 3.26, different other resistors and capacitors will be included to reach the goal proposed. Figure 3.27 presents the final design of the oTx with one white cold LED of 1 W.

Since the main "actor" of the oTx circuit is the LED, according to its characteristics the entire electronic circuit has to be designed. It is plotted first, as can be seen in Fig. 3.28.

As the optical receiver, the PIN PD is a blue-enhanced silicon PIN photodiode with a filter which removes sensitivity to unwanted infrared and a spectral range between 330 and 720 nm, the LED's wavelength is settled at 588 mm. Its maximum voltage is 3.3 V and maximum current is 350 mA. During operation, its current varies from 100 to 200 mA at 100 Ω R1.

The loading effect is compensated using an R1. This passive device also compensates the potential divider for varying tolerances in the resistors' construction.

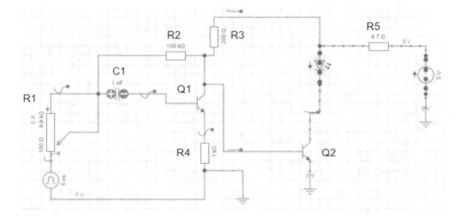

Fig. 3.27 Electronic circuits designed and simulated for the oTx driver

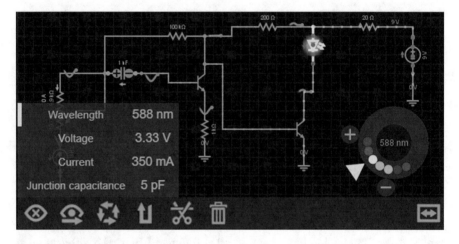

Fig. 3.28 LED's characteristics plotted

The current-limiting resistor (R5) is positioned in the circuit, in series with the cold white 1 W LED and the transistor's collector.

The base resistor has a large value relative to the current limiting resistor because this transistor amplifies base current by a factor of 45 (Fig. 3.29).

Since the simulation application does not allow changing the settings regarding the resistors' symbol according to the European one (the rectangle), the resistors' symbol displayed is the default one in the software ⋙ Fig. 3.30.

The list of all the passive and active elements used in the electronic circuit (Fig. 3.31) are:

LED—1 W, max 350 mA, and 3.3 V.
R1—10 KΩ variable resistor.
R2—resistor of 100 KΩ.

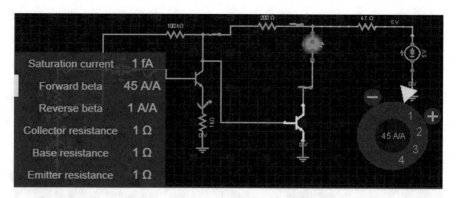

Fig. 3.29 Plotting Q2, the NPN transistor

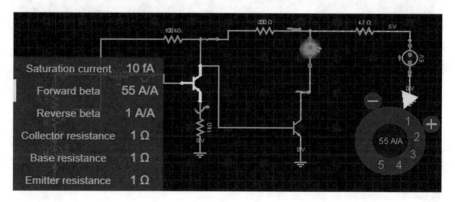

Fig. 3.30 Plotting Q1, the NPN transistor

R3—200 Ω.
R4—1 KΩ (10^3 Ω).
R5—4.7 Ω.
C1—polarized 1 nF (10^{-9} F).
Q1—2N3904 NPN transistor.
Q2—2SC5200 NPN transistor.
Vin—signal.
Vcc—5 V.

As noted in the 2 N3904 datasheet, the *DC current gain* h_{FE} may be at least 30 when $I_C = 100$ mA.

To ensure the transistor 2 N3904 saturation, the results from relation 3.24 have to exceed desired collector current I_C Pulse test for 2N3904 is as in.

$$tp \leq 300 \ \mu s \text{ and } \delta \leq 0.02 \tag{3.23}$$

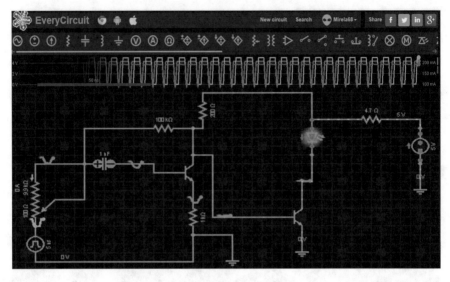

Fig. 3.31 Simulation of the electronic circuit of the oTx with EveryCircuit application

$$h_{FE} * \frac{V_{CC} - V_{BE}}{R_B} > I_C \qquad (3.24)$$

where:

Vcc—5 V.
V_{BE}—Base-emitter saturation voltage.

In Fig. 3.31, the final circuit with the waveform seen on the oscilloscope (at top of the window) is presented.

The waveform into specific points in the circuit is displayed as in Figs. 3.32–3.34).

1. *the input signal (Vin) colored yellow,*
2. *the input voltage (Vin) colored blue,*
3. *the output of LED, colored orange.*

Prior to purchasing hardware, when possible, designing, testing, and simulating the system is preferable to be done using one of many available useful applications.

In order to do this, *Proteus ISIS* application can be used that has some powerful tools, including more than 800 types of microcontroller, support many processor families along with lots of embedded peripherals being a reliable solution for circuit simulation and PCB professional design allowing to create a proper PCB for the VLC system.

Arduino programs can be written in Proteus Visual Designer using handy flow-charting methods and schematic Arduino shields that can be easily placed on schematic capture design space, then the entire Arduino system can be simulated, tested, and debugged in this software.

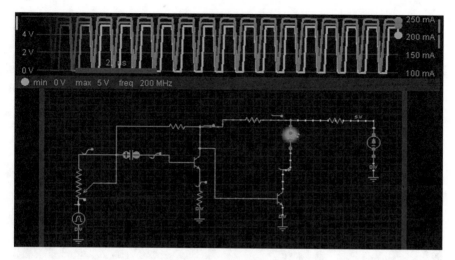

Fig. 3.32 The yellow waveform of the input signal highlighted

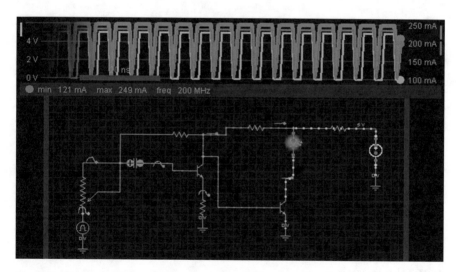

Fig. 3.33 The blue waveform of the input voltage highlighted

The Arduino Uno R3 is a board based on microcontroller ATmega328. The board has:

1. 14 digital input/output pins:

   (a) of which 3, 5, 6, 9, 10, and 11 can be used as PWM outputs,

2. 6 analog inputs
3. power jack,

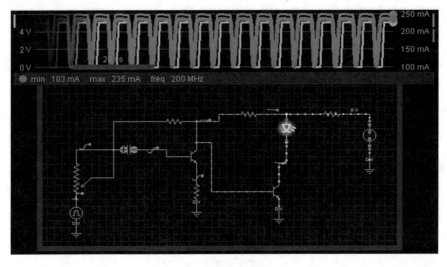

Fig. 3.34 The orange waveform of the LED highlighted

Table 3.6 Arduino Uno's main characteristics

Microcontroller	ATmega328
Operating voltage	5 V
Input voltage (recommended)	7–12 V
Input voltage (limits)	6–20 V
DC current per I/O pin	40 mA
DC current for 3.3 V pin	50 mA
Digital I/O pins	14 (where pin 6 provides PWM output)
Analog input pins	6
Clock speed	16 MHz
Flash memory	32 kB (ATmega328) of which 0.5 kB used by bootloader
SRAM	2 kB (ATmega328)
EEPROM	1 kB (ATmega328)

4. 16 MHz crystal oscillator
5. USB connection.
6. ICSP header.
7. reset button.

The technical characteristics (Table 3.6.) of Arduino Uno are essential for the entire oTx driver:

The oTx of the VLC system can be seen in Fig. 3.35.

When two different software are used for simulation, a Hex file is needed to program the microcontroller. First, the hex file has to be obtained in Arduino IDE. Therefore, prior to running the final oTx circuit in application, the code has been

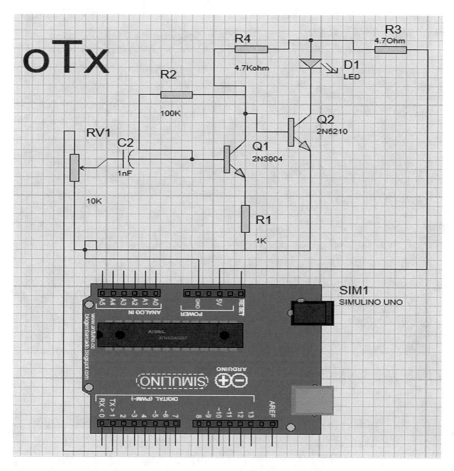

Fig. 3.35 oTx circuit with Arduino Uno (Proteus ISIS app)

written in Arduino software (authors used 1.8.5 version of Arduino) and prior to compilation, settings in Preference command are established (Fig. 3.35).

The file has to be saved, and then, in Arduino Preferences (File > > Preferences or Shortcut key: Ctrl + Comma), the verbose mode for compilation is enabled by ticking option as shown in Fig. 3.36.

Compiling the code, the HEX file path you will be displayed on the bottom of the window (Fig. 3.37).

After compilation, the hex file:

```
Send_EAN_8.ino.hex
```

will be added into the Proteus Arduino UNO board, as it can be seen in Fig. 3.37 and then, the oTx module can be tested by simulation (Fig. 3.38).

Fig. 3.36 Options in preference

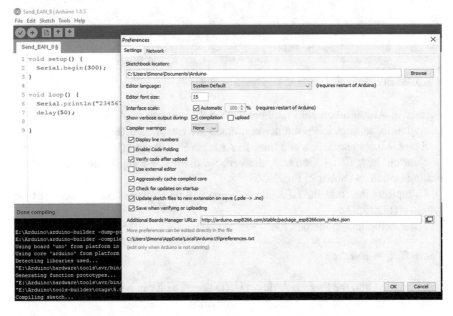

Fig. 3.37 Setting the compilation procedure in Preference (Arduino 1.8.5 app)

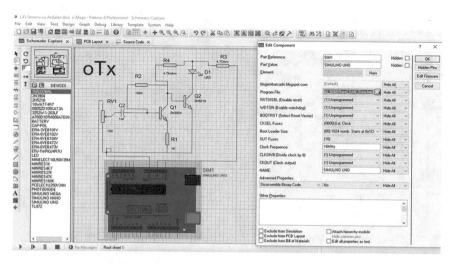

Fig. 3.38 Arduino Uno component edited in ISIS

The oRx Electrical Circuit and Simulation with Microcontroller

Data transmitted by light emitted by LED will be received by the PIN photodetector. The PIN PD has a specific output current that is converted into voltage. The level of output current is, generally, directly proportional with the level of illumination. Right after the PD, on the designed circuit, a Trans Impedance Amplifier (TIA) is required in order to overcome possible weak photocurrent signals. The longer the distance between the LED and PD, the weaker is the signal received by the PD.

When the current output of the PD is high, the high noise is also obtained, and therefore the signal received is proportionally damaged, hence an amplifier circuit is a suitable approach, even though the oRx will become more complex.

When choosing the suitable PD for the oRx design, both APD and PIN are taken into consideration since both photoconductive (APD) and photovoltaic (PIN) PDs have advantages and disadvantages.

The photoconductive PD has a reverse bias, high noise, is nonlinear, and generates dark current while the photovoltaic PD is linear, without bias, has low noise, and does not generate dark current.

As PIN PD has low-noise and precision characteristics, it is the preferable type of PD in the oRx circuit of the VLC system simulated aiming that the data signal emitted by LED to be received with low-level noise by PD.

Since the oTx module will send data through LED's light while the cap-lamp user moves on the main gallery under the oRx with the PIN PD embedded, the relative radiant sensitivity versus angular displacement is important to be taken into consideration in the proposed VLC positioning and monitoring system.

In this way, it is avoided designing any additional, expensive circuit to minimize noise and focus on the next stage of the oRx, the amplifier (Op Amp).

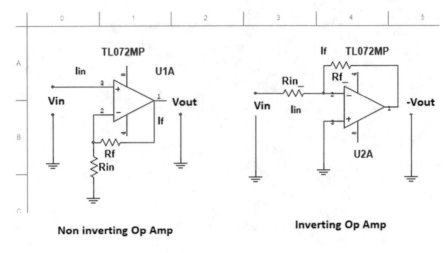

Fig. 3.39 Basic configuration of Op Amp

An Op Amp detects the difference between two voltage signals (applied to its two input terminals) and multiplies the signal with a Gain (A) (Eq. 3.25), often referred as "Open loop gain." The Op Amp can be controlled by connecting a resistive or reactive component, between one input and the output terminals.

Op Amps can have one of the two basic configurations, non-inverting and inverting, as seen in Fig. 3.39.

In case of the non-inverting Op Amp, the Gain is:

$$A = \frac{V_{out}}{V_{in}} = 1 + \frac{R_f}{R_i} \qquad (3.25)$$

In case of the inverting Op Amp, the Gain is:

$$A = -\frac{V_{out}}{V_{in}} = -\frac{R_f}{R_{in}} \qquad (3.26)$$

The *first circuit designed* for the oRx aims to test accurate sound communication in order to gain high distance and volume of the sound wave received by the speakers. This will act as a "witness" receiver during wireless communication accuracy tests.

The PIN PD used in the first electronic circuit simulation is a BPW20R produced by Osram company, with the characteristics specified in Table 3.7 and relative radiant sensitivity versus angular displacement as seen in Fig. 3.40.

The Op Amp-type TL072 used in this design and simulation is manufactured by Texas Instrument (TI) Inc. Authors use the *Multisim* application since this software has many specific tools for simulation before prototyping and the circuit files are also available.

Table 3.7 Characteristics of BPW20R PIN PD

Parameters	Values
BPW20R	
Wavelength ($\lambda_{0.5}$)	550–1040 nm
Forward voltage for $I_F = 50$ mA	1.3 V
Short current (I_{SC})	7.4 μA (1000 lux)
Radiant sensitive area (mm²)	7.5 mm²
Rise time (tr)	3.4 μs
Fall time (tf)	3.7 μs
Reverse voltage	10 V_{DC} (max.)
Half angle (ϕ)	±50°
Dark current (I_R)	2 nA
Capacitance (C) $V_R = 0$ V, $f = 1$ MHz	1.2 nF
Capacitance (C) $V_R = 5$ V, $f = 1$ MHz	400 pF

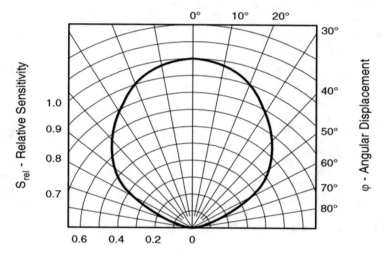

Fig. 3.40 Relative radiant sensitivity versus angular displacement

TL072 is a JFET-input Op Amp that has low input bias and offset currents and fast slew rate, with low harmonic distortion and low noise.

The most important features of the Op Amp taken into consideration for the first circuit are the Gain bandwidth product (GBWP) (3 MHz), maximum power supply (18 V_{DC}), the number of channels (Ch.) (2), and the maximum output voltage (Vout$_{Max.}$) ($V_{DC} = 12$), for high band with a cutoff frequency $f_c = 3$ MHz.

Main characteristics of TL072 amplifier (that we considered in case of this project to reach our goal) are presented in Table 3.8.

Limiting the key factors, as high data rate and high distance between oTx and oRx in the setup, also limit the bandwidths of the PIN PD and the use of a simple operational amplifier and therefore a low-cost circuit. Simulation of the oRx with Op-Amp TL072 is presented in Fig. 3.41.

Table 3.8 TL072 parameters

Parameters	Values
OP-AMP type	TL072
Large signal voltage amplification	50–200 V/mV
Rise and fall time	1.5 ns
Gain bandwidth product (GBWP)	140 kHz ($V_{out} = \pm 10$)
Voltage noise density (e_n)	8 nV/$\sqrt{Hz}$ (30 Hz)
Current noise density (i_n)	2.7 fA/$\sqrt{Hz}$ (30 Hz)
Slew rate	290 V/us
OP-AMP saturation ($R_L = 100 \, \Omega$)	± 10 V
Supply current (I_{CC})	2.5 mA
Input offset current	5–100 pA
Input bias current	65–200 pA

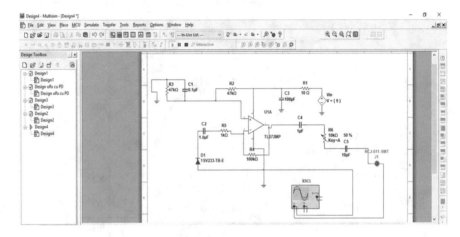

Fig. 3.41 Simulation of the oRx with Op-Amp TL072

We have tested *the second circuit* of the oRx (Fig. 3.42) to gain high accuracy of data transmitted at the highest distance (between oTx and oRx) with the best attainable BER.

The PIN PD used into the electronic circuit simulation is a planar silicon photodiode into a recessed ceramic package that incorporates an infrared rejection filter. The diode has very high shunt resistance and a good blue response.

The electro-optical characteristics of the VTB8440B PIN PD are specified in Table 3.9.

The Op Amp type NE5532P is used in this second design and simulation. It is also used *Multisim* application, to simulate the electronic circuit.

NE5532P is a dual Op Amp with low noise and high performance. NE5532P has improved output drive capability and small signal with power bandwidths.

The most important features of the NE5532P Op-Amp are presented in Table 3.10.

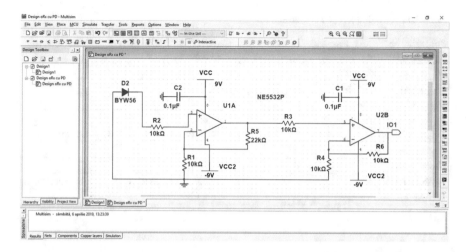

Fig. 3.42 Simulation of the oRx electronic circuit in *Multisim* with Op-Amp NE5532P

Table 3.9 PIN PD VTB8440B

Parameters	Symbol	Values
VTB8440B		
Spectral application range	λrange	330–720 nm
Open circuit voltage H = 100 fC, 2850 K	Voc	420 mV
Peak of spectral response	λp	580 nm
Dark current	I_D	2000 pA
Radiant sensitive area	A	5.16 mm^2
Junction capacitance	C_J	1 nF
Angular response (degrees)	$\Theta_{1/2}$	±50
Breakdown voltage	V_{BR}	10 V_{DC} (max.)
Specific detectivity	D*	2.2 × 10^{12} cm √Hz/W
Noise equivalent power	NEP	1.1 × 10^{-13} W/√Hz

Table 3.10 Op Amp NE5532

Features	Values
Large signal voltage range	±3 to ±20 V
DC voltage gain	50,000
Large signal voltage gain	50–100 V/mV
Power bandwidth	140 kHz
Input noise voltage	5.0 nV/√Hz
Slew rate	9 V/μs
Supply current (I_{CC})	2.5 mA
Small signal bandwidth	10 MHz

Table 3.11 Arduino Mega's characteristics

Microcontroller	ATmega2560
Operating voltage	5 V
DC current per I/O pin	40 mA
DC current for 3.3 V pin	50 mA
Input voltage (recommended)	7–12 V
Input voltage (limits)	6–20 V
Digital I/O pins	54 (of which 14 provide PWM output)
Analog input pins	16
Clock speed	16 MHz
Flash memory	256 kB of which 8 kB used by bootloader
SRAM	8 kB
EEPROM	4 kB

The voltage operational amplifier circuit is placed between the PIN PD and the microcontroller embedded on the Arduino board.

For simulation the Arduino Mega board based on microcontroller ATmega 2560 is used.

The board has:

1. 54 digital input/output pins

    1. of which 14 can be used as PWM outputs,

2. 16 analog inputs
3. a power jack,
4. a 16 MHz crystal oscillator,
5. 4 UARTs (hardware serial ports)
6. USB connection.
7. an ICSP header.

Universal Asynchronous Reception and Transmission (UART)—communication protocol permits Arduino board to serially communicate with other devices. UART system communicates with digital pin 0 (R_X), digital pin 1 (T_X), and with another computer via the Universal Serial Bus (USB) port.

The In-Circuit Serial Programming (ICSP) is a protocol used to programme microcontrollers and pins are used to update the firmware or reinstall a bootloader.

Technical characteristics (Table 3.11) of Arduino Mega are essential for the entire oTx driver:

More performant microcontrollers with more advanced functions can be used in order to upgrade the system performance that is simulated here [19].

With the same two different software used for simulation, the Hex file is also needed to program the microcontroller. First, the hex file is obtained in Arduino IDE. Therefore, prior to running the final oTx circuit in application, the code has been written in Arduino software and prior to compilation, settings in Preference

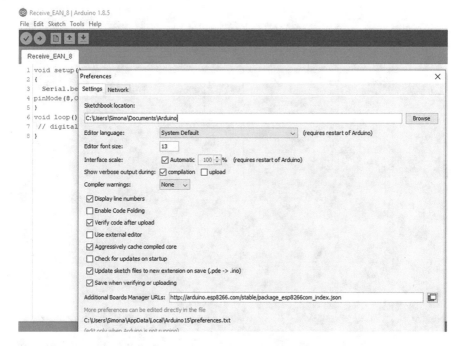

Fig. 3.43 Verbose output during compilation display

command are established. The file is saved, and then, in Arduino the verbose mode for compilation is enabled by ticking option as shown in Fig. 3.43. Simulation in *Proteus ISIS* app of oRx with Arduino Mega and TL072 are presented in Figs. 3.44 and 3.45, and settings with simulation of the oRx circuit with Arduino Mega and NE5532P is presented in Fig. 3.46.

Channel Model and its Optical Impulse Response

The effort of designing the complete VLC system for local wireless communication in an underground mine is a challenging one. Not only the oTx and oRx have to be properly designed but the environmental conditions and the optical medium have to be carefully considered. Therefore, the effects of lights' dispersion, the optical signal attenuation, and scattering due to polluted environment, have to be all taken into consideration to avoid a significant limitation of the system's performance. An appropriate oTx driver setup with proper optics, as well as proper optics in front of PD can significantly mitigate the effects of optical beam dispersion and optical signal loss.

An accurate visible light communication channel model applied to the underground mine, or at least as close as possible to a real model of the optical channel

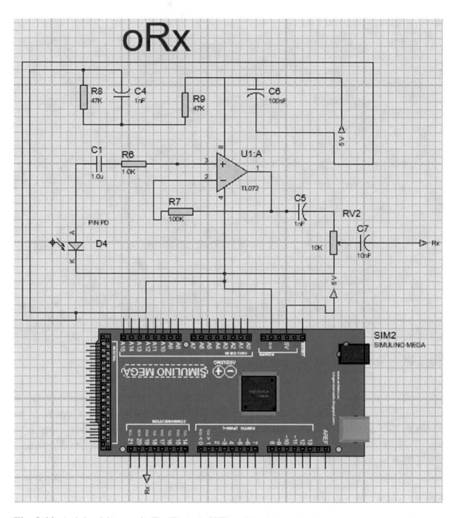

Fig. 3.44 Arduino Mega and oTx (*Proteus ISIS* app)

with all its particular characteristics is also critical to be well defined and developed when we design a wireless transmission system based on VLC.

The optical channel for visible light is a time-invariant, linear, and memoryless system and its impulse response has a finite duration [20].

When high data rates have to be achieved, in case that the signal bandwidth goes outside of the channel coherence high and low bandwidth limits, the optical signal suffers from slow fading. In this case, the optical channel can be modeled as frequency selective due to light's dispersion.

The light distribution indoor has been already simulated on computer for different topologies (LoS or NLoS) and many ray tracing algorithms (such as Monte Carlo Ray Tracing—MTRC, modified MCRT, or deterministic algorithms, such as Barry's

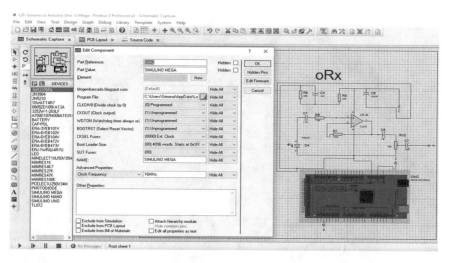

Fig. 3.45   Settings for simulation the oRx circuit with TL072

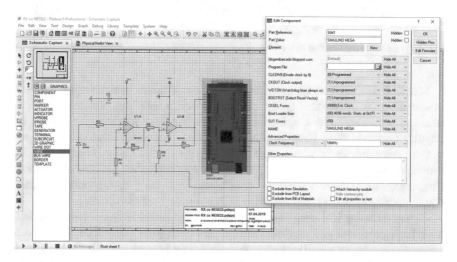

Fig. 3.46   VLC setup for a LoS topology underground

algorithm, etc.) have been tested and demonstrated. Some of the most important findings when investigating the visible-light wireless communication channel are the RMS DS, Path Loss (PL), BER/SNR. Using MCRT algorithm for a clear demonstration of the ray distribution (both in LoS and NLoS topologies) the results showed that the PL (in dB) is linear over logarithmic distance and (in both LoS and NLoS topologies) its range starts at -27 dB and ends at -80 dB. RMS DS has been proved to be between 1.3 and 12 ns for LoS links and between 7 and 13 ns for NLoS links [20].

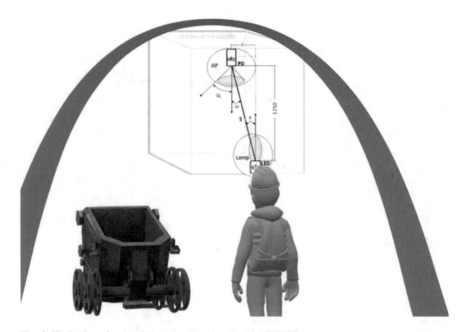

Fig. 3.47 Settings for simulation the oRx circuit with NE5532P

Monte Carlo ray-tracing algorithm allows an accurate evaluation of CIR for environments with complex geometries and indoors where high numbers of reflections are necessary to be considered. IR and visible light wireless channel and its impulse response have been already investigated indoor for isotropic (clean air without any particles in suspension) indoor environments. Many attempts for CIR evaluation under noisy (bright sun) and harsh meteorological conditions outdoors (fog, rain, snow) have also been investigated and the remote optical wireless communication has been demonstrated with promising results [21, 22].

Computer numerical simulations, in spite of high time and resources consumed, have been presented and mathematical models have been established, taking into account one, two, or a finite number of straight light rays, or rays that bounce from the obstacles indoor, such as ceiling, floor, or different static or moving obstacle inside a room with furniture with different geometries, colors, and type of surfaces.

In Fig. 3.47, the geometry of an LoS topology with light propagation model in a local wireless communication system in visible light is represented (with incident and refracted rays), between the miner's helmet cap-lamp and the illumination infrastructure.

According to the second chapter of this book, where the mathematical model for the indoor visible light channel model has been presented, the key characteristics considered for the optical channel model simulated in case of the underground VLC setup proposed, are presented in Table 3.12.

Table 3.12 Characteristics considered for the optical channel [23]

Nr.	Characteristic	Symbol	M.U.	Value
1	Semi-angle of half power (see Fig. 2.12)	φ	^{0}	45°
3	Transmitted optical power by LED	Ptot	Lm	80
4	Active area of PD	A_{PD}	mm^2	5.15
5	Refractive index of a lens at PD	Index	–	1.55
6	FoV of the PD	FoV	Degrees	45
7	Dimensions of the underground considered (see Fig. 3.47)	$L \times l \times h$	m	$2.2 \times 2 \times 3$
8	Distance between LED and PD	d	m	1.470
9	Position of PD related to LED	X_T, Y_T	m	2, 1.25

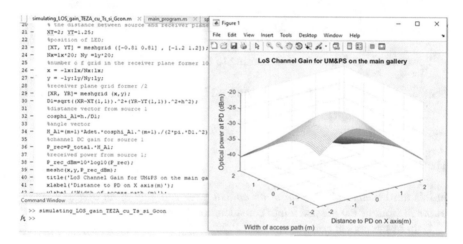

Fig. 3.48 Results of the first scenario simulated

First Scenario of Simulation

In this first scenario, neither **gain** of an optical **filter** nor **gain** of an optical **concentrator** in front of PD is taken into consideration (see Fig. 3.48).

The Lambertian order of emission is:

$$m = -\log 10(2) / \log 10(\cosd(fi)) \tag{3.27}$$

Channel DC gain from source (according to equation in chap. 2), in this first scenario considered, is:

$$\text{H\_LoS} = (m+1) * \text{APD.cosphi\_LoS.}^{(m+1)} / (2 * \text{pi.} * d1.^2) \tag{3.28}$$

Power received from source, according to Eq. 2.38, in this first scenario, is:

$$P\_rec = P\_total. * H\_LoS \qquad (3.29)$$

The Second Scenario of Simulation

In this second scenario *gain of an optical filter* is taken into consideration but *gain of an optical concentrator* in front of PD is not taken into consideration (see Fig. 3.49).

Ts—optical transmission of the band-pass filter.

Channel DC gain from source, for this second scenario considered is:

$$H\_LoS = Ts * (m + 1) * APD.cosphi\_LoS.^{(m+1)}./(2 * pi. * d1.^2) \qquad (3.30)$$

Power received from source, according to Eq. 2.38 in this first scenario, is:

$$P\_rec = P\_total. * H\_LoS * Ts \qquad (3.31)$$

The Third Scenario of Simulation

In this third scenario both *gain of an optical filter* and *gain of an optical **concentrator*** in front of PD are taken into consideration (Fig. 3.50).

Gain of the optical concentrator is:

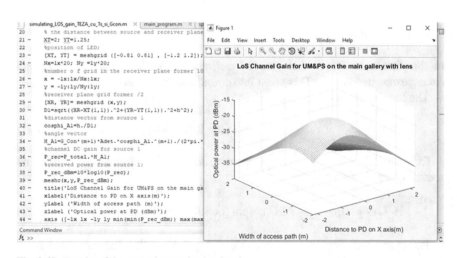

Fig. 3.49 Results of the second scenario simulated

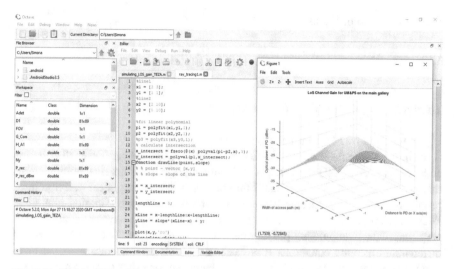

Fig. 3.50 Results of the third scenario simulated in Octave

$$G\_Con = \left(\text{index.}^2\right)/\sin\left(\text{FoV}\right) \tag{3.32}$$

Channel DC gain from source, for this second scenario considered is:

$$H\_LoS = G\_con * Ts * (m+1)$$
$$* \text{APD.cosphi\_LoS.}^{(m+1)}./\left(2 * \text{pi.} * \text{d1.}^2\right) \tag{3.33}$$

Power received from source is:

$$P\_rec = P\_total. * H\_LoS * Ts. * G\_Con \tag{3.34}$$

As can be seen, after the three scenarios simulated in Octave, according to mathematical models presented in Chap. 2, the optical power is first improved due to the lens in front of LED (Fig. 3.49) and then with optical filter and optical concentrator added in front of PD, as can be seen in Fig. 3.50.

All figures above, from 3.27–3.34, are presented as written in Octave editor.

Following simulation in three different scenarios, the channel gain is as follows:

First Scenario of Simulation (Fig. 3.46)

Power received by PD from LED is about: -32 dBm

Second Scenario of Simulation (Fig. 3.48)

Power received by PD from LED is about: -27 dBm

Third Scenario of Simulation (Fig. 3.49)

Power received by PD from LED is about: -22 dBm

In order to simulate, in three different situations, the optical power distribution and CIR, we used GNU Octave, a high-level language, for numerical computations that provide a convenient command line interface for solving linear and nonlinear problems numerically. Octave has extensive tools for solving common numerical linear algebra problems, finding the roots of nonlinear equations, integrating ordinary functions, manipulating polynomials, and integrating ordinary differential and differential-algebraic equations. It is easily extensible and customizable via user-defined functions written in Octave's own language, or using dynamically loaded modules written in C++, C, Fortran, or other languages [24].

As a first conclusion, for the same LoS topology in a similar FoV scenario, considering the same environment and also unchanged position of both LED and PD, with optical filter and lens added in front of the PD, the optical power gained is definitely improved.

A complex analysis from the channel's optical behavior point of view has to be done prior to deploying a VLC system for local wireless data transmission since the underground environment has some benefits as well as drawbacks.

Benefits

The lack of natural light and a tiny level of artificial light from different other sources (as lighting fixtures from the illumination infrastructure or the mining equipment) is an advantage underground, since the AWGN that has to be considered, is very low.

The underground link length is relatively short because the distance between the miners' helmet cap-lamp and the illumination network is also short.

Position and inclination of the oTx relative to oRx can be clearly determined to set up an LoS topology with the shortest distance between LED and PD with a well-known elevation angle.

The predominant color underground is black and gray, the light absorption being very high, the NLoS ray acquisition is negligible and the few ray bounces and their low energy do not influence the signal strength of the main optical beam light.

Most of the materials underground are metal, wood, or rock with top roughness surfaces, therefore just a few light' bounces can be considered and tested for an NLoS optical gain with the specific refractive indexes [25].

Drawbacks

The primary restraint of this VLC system for UO&M is to keep the scattered light beam from LED, wide enough to illuminate the worker's path and as much as narrow to focus the ray of light onto the oRx and preserve as much as possible from the data stream sent. The optimal LEDs' inclination has to be determined as to illuminate the workers' path and send useful data to the roof of the gallery where the oRxs are placed, embedded into the illumination network or into the optical fiber that acts like the backbone of the communication system already installed into the underground of most of the modern mining companies.

Another important drawback, difficult to be proper estimated in this moment, is the velocity, density, individual configuration (geometrical irregular shape), and composition (mixture of dust, coal, moisture) of the suspended particles in the optical path of the light beam that scatter and absorb the light wave and decrease the optical beam energy.

The optical path medium is highly influenced by this polluted environment with tiny suspended particles of coal and rock with different dimensions, shape, and grade of humidity; therefore, light scattering and absorption, as well as extinction coefficient, are difficult to be calculated based on present mathematical models [25], difficult to estimate or simulate with numerical methods and powerful machines.

While simple analytic expressions exist for the absorption and scattering properties of dust grains which are either very small or very large compared to the wavelength of the incident radiation, however, suspended and mixt particles of coal and rock produced by machinery underground during operation, have nonspherical, irregular shapes.

The relation between incident light intensity and transmission light intensity can be expressed as follows [26]:

$$I = \left(I_\varphi, I_\omega\right) e^{-1.5 \frac{CDK_e}{d}} \tag{3.35}$$

where:

C—concentration of the suspended particles in the unit volume ($/cm^3$)
D—light path/distance between oTx and oRx (cm)
φ—incidence angle
ω—refracted angle
K_e—the light extinction coefficient

$$k_e = Q_{sca} + Q_{abs} = \frac{total\ energy\ scattered\ and\ absorbed\ per\ unit\ time}{incident\ energy\ per\ unit\ area\ per\ unit\ time} \tag{3.36}$$

where:

Q_{sca}—light-scattering coefficient
Q_{abs}—light absorption coefficient

The extinction coefficient (k_e), due to many variable underground, cannot be accurately calculated according to Eq. 3.37:

$$k_e = \lambda \frac{ln \frac{I_r}{I_i}}{4\pi m_t} \tag{3.37}$$

where:

λ—wavelength
I_r—intensity of light at destination
I_i—intensity of light at source
m_t—medium thickness

Therefore, numerical simulations and practical tests have to be conducted to determine the extinction coefficient for different scenarios in the underground polluted environment both far and in the proximity of the working spaces where machinery operates, in order to estimate the proper work of the VLC system design and develop an adaptive system. The proper design of the optics, both in front of LED and in front of the PIN PD, is clearly very important, from this point of view, as well.

3.3 System Implementation

3.3.1 Hardware Implementation

We have established a general frame with a specific set of the system requirements at the beginning of the project. We also completed the simulation of both transmission module and the reception one. The oTx module consisting of the electrical board, the electronic PCB with microcontroller, the light emitter, and its optical system as well as the oRx module with the electrical board with TIA, the electronic PCB with microcontroller, the photodetector and the appropriate filter and optics have been also completed. The necessary key characteristics for a solid, reliable local wireless VLC transmission for the UP&MS proposed have been also highlighted. We have also analyzed the underground characteristics with the polluted environment and optical medium with their key properties as well as the multiple effects of the suspended particles on the optical signal in an LoS topology.

oTx Module

The miner's cap lamp is projected to have two functions: illumination and data communication. The primary functionality that has to be considered is illumination. Wireless data communication is the secondary function of the miner's cap lamp;

therefore, design and prototyping have to take into consideration these both functions, each with a proper, balanced importance.

According to records and conclusions of the Mine Safety and Health Responsible Bodies worldwide, because the quality of the illumination is low, many accidents happen underground during shifts. Lighting, especially from the miners' helmet cap-lamp, plays a critical role for miners as they visually inspect the mine roof, ribs, back, and floor on their moving path to avoid hazards. Objects on this path associated with these hazards are typically of very low contrast and reflectivity. There are also age-related factors that require a better quality of light. The night vision of the elder workers is also reduced because there are changes in the eye that include decreased pupil size, cloudier lens, and fewer rod photoreceptors that are very sensitive to light [27].

When the proper optical source for the system designed has to be selected, a comparison between most of the off-the-shelf LEDs, and semiconductor LDs (SLDs), with their key characteristics, is not only useful but necessary.

The optical spectral width of a LED is between 25 and 100 nm and the one of an SLD is situated between 0.01 and 5 nm. As for the electrical to optical conversion, the LED's efficiency is not more than 20% when the SLD's efficiency is between 30 and 70%.

LEDs are low temperature-dependent while SLD is high temperature-dependent. The drive and control circuitry for LED are simple to use and control while SLDs need a threshold and temperature compensation circuitry. LEDs experience higher harmonic distortions than SLDs. When LEDs are used as emitters, at the oRx filter is wide, increasing the noise and in case of SLDs, the filter is narrow, thus the additive noise is low. Finally, the most off-the-shelf LEDs have low cost while SLDs are more expensive [23].

There are few key characteristics of LEDs as benefits but there are also some drawbacks that have to be taken into consideration in order to select the most suitable LED. So, evaluating all these key characteristics, trade-offs are necessary to be done, here also.

LEDs have to flicker (blink—ON/OFF) with a frequency not visible by eye (preferable higher than 1 kHz).

The frequency response of the off-the-shelf white LEDs is not listed by manufacturers thus making the selection and verification hard to implement. And, further amplification on the oTx is certainly possible for the brightness with a correct electrical design [28].

White LEDs with a yellow phosphor layer usually have a slow response and the cut-off frequency between 1 and 5 MHz.

Off-the-shelf standard LEDs (red, blue, green, white, IR, UV) have cutoff frequency between 10 and 50 MHz (sine wave). The cutoff frequency is the maximum frequency at which LED drops at half level from its initial light intensity.

The rise time of LED is highly affected by its maximum intensity. On the other hand, low intensity increases the rise time but a low intensity shortens the link length and therefore the VLC setup efficiency. Then again, high intensity needs more power. More power is not allowed in lighting systems underground.

Fig. 3.51 LEDs panel

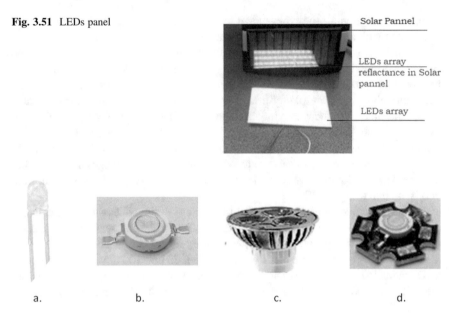

Fig. 3.52 Different types of LEDs tested

The ability to use a white LED as a modulating RF transmitter is from the fact that the LEDs modulation bandwidth is capable of a few MHz and after some equalization techniques over 30 MHz [29].

The LED in a VLC system is a major source of nonlinearity that is important when analog OFDM modulating signal is used because it degrades the bipolar time domain DCO-OFDM signal through amplitude distortion, clipping of the lower peaks, and clipping of the upper peaks. Selecting an LED with high AC/pulsed current level enhances the performance. Considering an LED with low voltage-current slope characteristics, the error performance is improved [30].

Before the setup of the oTx presented here is final, many types of LEDs, array of LEDs, PIN PDs, and solar panels have been first tested with a prototype designed and manufactured with the aim to send sound. This was the "witness" prototype for the final designed prototype, the one for wireless data communication.

Some LED arrays (as, e.g., the one in Fig. 3.51) have been tested with the sound transmission system, used as the "witness" prototype of the wireless communication through visible light.

There have been tested may types of LEDs. Some of them, presented in Fig. 3.52 such as VLHW4100 from VISHAY (a), LL-HP60MW1L-S (b) from LUCKY LIGHT, P001L4Z11 from TDS (with three LEDs embedded), and VLHW4100 (d) from VISHAY and array of LEDs.

The LED used in the wireless VLC prototype is an SMD one placed on a cooler (VLHW4100) (see Figure 3.52d and 3.53).

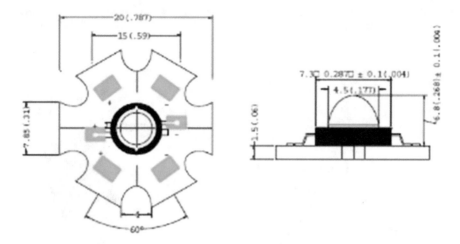

Fig. 3.53 LED's dimensions

The circuit of the oTx with the 1 W white cold LED and component track lead capacitance and inductance have to be kept to a minimum, to be able to drive the LED up to 1 MHz.

This LED used with such circuit can be a feasible choice without other complex pulse-shaping oTx drive circuits or with very short lead lengths.

The LED has 4.5 mm diameter and its transparent cover is 4.6 mm high. These dimensions are important when determining the optimum distance between the LED and lens in front of the LED.

The LED's key characteristics embedded in the present VLC prototype are presented below.

1. V_F: min 3.0 V max 3.6 V.
2. Luminous Intensity: 90–110 lumen.
3. Max current 350 mA.
4. CCT (color temperature): between 3200 and 3500 K.

Beside LED, many other parts of the designed system have to be taken into consideration, together with the performance and limitation of the designed electronic circuit as well as the microcontroller used that is embedded in the Arduino Uno board. The board has a maximum allowable voltage of 5 V, maximum current output of 1 A, and can send a square wave with the maximum frequency of 50 kHz. ATmega328p type of microcontroller has been used for prototyping and testing the oTx setup designed for the UP&MS.

Transistors used for designing and prototyping the oTx of the VLC system are 2 N3904 and 2SC5200 as presented in Fig. 3.54.

The electronic circuit of oTx temporary fixed on breadboards can be seen in Figure 3.55a, b.

The same oTx circuit tested with LED is presented in Fig. 3.56.

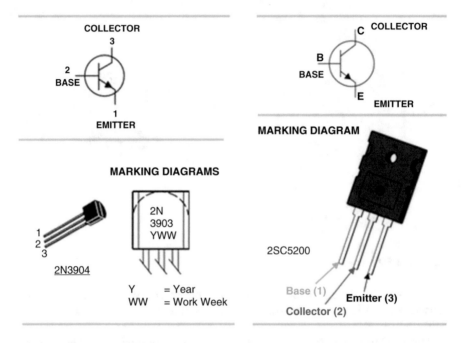

Fig. 3.54 Transistors 2 N3904 and 2SC5200

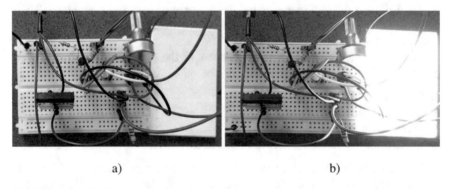

a)   b)

Fig. 3.55 oTx with LED array panel

The oTx circuit on board tested with LED is presented in Fig. 3.57.

To keep the systems' cost low, the Arduino Uno R3 board with microcontroller ATmega328 is also used for the oTx for testing purpose.

Fig. 3.56 oTx breadboard
with LED

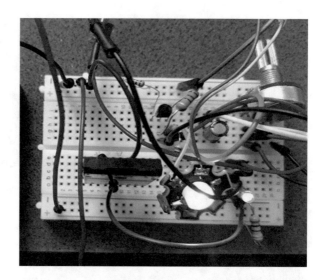

Fig. 3.57 oTx board

oRx Module

One of the most important components of the entire design is the PD. Unlike the
LED, a PIN photodiode must be reversed biased in order to function properly. Most
PDs require a significant amount of voltage. In this scenario, there are several
approaches, the first involves testing a number of solar panels and then PDs that
have a very broad acceptance range, making it suitable for any kind of LED emitter.
The second step was to test a number of PDs and pick a PD that is very narrow in its
range with low cost and acceptable performance.

A number of panels with photocells (solar panels) were under test to check the
distance and accuracy of the sound in the first designed system. However, for a high
accurate VLC system for data transmission, they are not suitable since they are not
high-frequency light sensing. Both the rise and fall time are too long (order of tens to
hundreds of milliseconds).

Fig. 3.58 Types of photocell panels tested

A diode is a two-terminal component that has low resistance to current flowing in one direction and high resistance to current in the opposite direction. A silicon diode starts to conduct current when the forward voltage across it exceeds the threshold of about 0.6–0.7 V.

Most of the off-the-shelf photocells spreadsheets indicate 60 ms rise time and 25 ms fall time, therefore, the highest frequency they can handle is below 11 H, meaning these values for high data rates necessary for the VLC system designed are not the targeted ones.

Datasheets of the PIN PDs indicate the rise and fall times. For example, most of the regular PDs, for 100 ns rise/fall times (each) would tolerate maximum of 5 MHz signaling. So, keeping the value into a safety path, 1 MHz would be adequate for a—not very performant, but a functional, reliable, and low cost—VLC system.

On the other hand, both photodiodes and phototransistors are favorite options for sensing high-speed light pulses at low to moderate intensity (when we consider a short or medium–long distance from the LED source).

For high signaling speeds and lower signal intensity, from cheap to quite expensive high-speed Silicon Avalanche Photodiodes (Si APD) are available on the market (as, e.g., APD C30902—about 50$ or PDA36A—above 900$).

Si APD C30902 series (Silicon and InGaAs APD Preamplifier Modules) have fall time and rise time of 0.5 ns, allowing a 1 GHz signal and with a light entry angle of 130°.

Although way beyond the PIN PD's key characteristics, the APD additively needs a high voltage power supply to compensate the large temperature variations that would greatly impose on its constant responsivity over temperature.

PDA15A–Si is a switchable Gain Detector with wavelength from 350 to 1100 nm, 10 MHz BW, the active area of 0.018 mm^2, impulse response of 1 ns (FWHM—full width at half maximum). Such improved values of fall and rise time, though, are not useful since they are way beyond the LED can support.

The optical power of the signal sent does not affect the dimension of the PD's sensitive area. The noise, on the other hand, is inversely proportional to the square root of the PD's sensitive area. The ratio between signal and noise is proportional to the square root of the PD's photosensitive area.

Fig. 3.59   Different types of PD tested

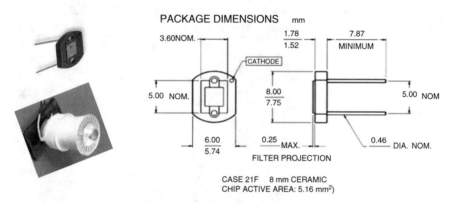

Fig. 3.60   PIN PD used for the oRx in the VLC system

As a conclusion, when the emitted signal intensity from LED is high enough, using suitable lenses, with a signal bandwidth not very high, then an acceptable VLC system can be designed and manufactured with signaling frequencies of hundreds of kilohertz, or even up to megahertz depending on the specific LED's characteristics used as emitter.

Many types of photocell panels have been tested, two of them as seen in Fig. 3.58.

There have also been tested many types of PIN PDs and APDs, some of them presented in Fig. 3.59 (BPW34 (a), VEMD5510C (b) from VISHAY, VTB8440BH (c) and VTB8440BH PHOTODIODE, IR FILTERED 85414090 (d).

Since the optimum choice of LEDs are those with short wavelengths due to their advantages (improved peripheral motion detection, reduce disability glare for older workers, and avoid easily any floor hazard due to early detection), PDs have to be able to detect light with this wavelength.

The most suitable choice from all the alternatives presented above is PIN PD VTB8440BH Photodiode, IR Filtered 85,414,090 as can be seen in Fig. 3.60.

An operational amplifier (Op-Amp), used mostly in analog electronics, is an integrated circuit that amplifies the difference between two input voltages and produces a single output. Fundamentally is a voltage amplifying device designed to be used with external components (such as resistors and capacitors) between its output and input terminals. These feedback components determine the resulting function or

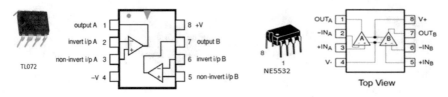

Fig. 3.61 Op Amps tested and used in VLC system

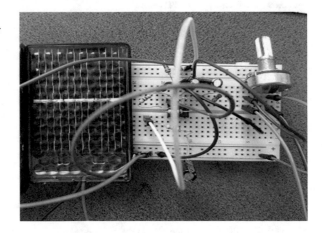

Fig. 3.62 oRx circuit on breadboard tested with solar panel

"operation" of the amplifier and by virtue of the different feedback configurations whether resistive, capacitive, or both, the amplifier can perform a variety of different operations [31].

The proposed oRx circuit uses a TIA, which converts current (from the photodiode) into voltage, and then amplifies the signal.

There are many amplifiers on the market and also a broad literature that compares the available amplifiers on the market taking into account their key characteristics applicable for a functional VLC system [32].

The key characteristics of an amplifier to consider for a suitable VLC system are:

1. the input capacitance (Cin) from (Cdiff + Ccomm) [pF],
2. rise and fall time [ns],
3. gain bandwidth product (GBWP) [MHz],
4. voltage noise density (e_n) [nV/√Hz] (>100 kHz),
5. current noise density (i_n) [fA/√Hz] (>100 kHz),
6. slew rate [V/us];
7. saturation (RL = 100 Ω) ± [V].

The GBWP must be at least 50–60% of the value specified on the op amp's datasheet [33].

Op Amps tested for the first circuits are LM386 and TL072 and the one finally choose for the VLC system in the oRx module is NE5532 (Fig. 3.61).

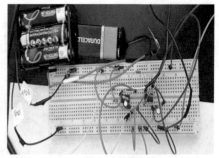

Fig. 3.63 Both oRx circuits on breadboards

Fig. 3.64 Both oRx circuits on boards

The first two (LM 386 and TL072) were used for the sound communication and the third one (NE5532) has been used for the oRx in the VLC prototype.

The system designed aims to send data to a distance between 1200 and 1450 mm in underground spaces with clean as well as polluted environment (particles of coal and rock suspended in the air) that is to be found close to the working room due to material cutting. The string of data to be sent is quite short and simple (EAN 8 code) embedded into the miner's cap lamp as its own ID.

The first circuit with solar panel for the oRx is presented in Fig. 3.62.

The second circuit with PIN PD for the oRx is presented in Fig. 3.63.

The final circuit on board with PIN PD for the oRx is presented in Fig. 3.64.

The Arduino Mega board based on microcontroller ATmega 2560 is used for the oRx.

The LED of the lamp continuously sends its own code (ID lamp) when the personnel that wear the cap-lamp on her/his helmet is moving through the main gallery. Therefore, the final prototype that aims to test and prove the concept has the oTx on wheels to be able to be moved during tests.

The entire prototype realized for data communication is presented in Fig. 3.65. The oTx module (left in Fig. 3.65) has the LED-type VLHW4100 from Vishay and oRx module (right in Fig. 3.65) has the PIN PD type VTB8440BH with IR filtered.

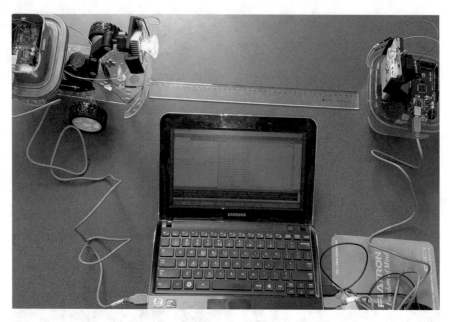

Fig. 3.65 The final prototype (left oTx and right oRx) under test

Both modules are aligned in an LoS topology, positioned at about 500 mm apart and the oTx is moved further during the test to determine the longest distance possible achieved for data accuracy communication.

Data sent/received during tests is displayed on the laptop's screen.

During tests, with this raw, low-cost manufactured prototype, the greatest distance achieved between the oTx and oRx, while communication is still possible, is 400 mm.

The next step for the prototyped system is to realize both circuits (oTx and oRx) on PCBs. This phase must be carefully developed since any designing that is incorrect can make circuits unstable because of the parasitic inductance and capacitance generated at high frequency by the PCB traces [32].

3.3.2 Software Implementation

The UP&MS consists of a hybrid communication network where the local wireless transmission is based on visible light and the remote transmission relies on a cabled network.

In this section, the focus is on the wireless local transmission through visible light. The algorithm used with flowcharts, frames, and packets is going to be presented in this work from now on.

Fig. 3.66 Odd, even, and CRC positions of the digits on the bar code, an example

The EAN-8 barcode that is stored into the oTx is continuously sent by the miners' helmet cap-lamp while the workers are underground and the lamp is ON. Although the checksum (CS) algorithm does not verify the authenticity of data received, the checksum used here as the EAN-8 barcode, aims to verify the data integrity.

This is the best choice considered in order to keep the system simple and efficient.

In order to verify the correct barcode, the algorithm to check data integrity, therefore any error (presented in the Fig. 3.66 as example), is the following:

STEP 1—All digits situated on odd (O) positions are added together:

$$2 + 4 + 6 + 8 = 20$$

STEP 2—The result is multiplied by 3:

$$20 \times 3 = 60$$

STEP 3—All digits situated on even (E) positions are added together:

$$3 + 5 + 7 = 15$$

STEP 4—Digits resulted in steps 2 and 3 are added together:

$$60 + 15 = 75.$$

Since the control digit is a result of a modulo 10 calculations, CS obtained is the difference between step 4 and the next tenth number:

$$75 + CS = 80$$

therefore

$$CS = 80 - 75 = 5.$$

The CS algorithms, as the one in EAN-8 is used, are techniques that handle errors with feedback. A CS is a many-to-one mapping of a large amount of data into a sum, as for 8, 16, or 32 bits.

Usually, in most systems, the transmitter (T) and receiver (R) calculate CS of the data in the packet. The T calculates the CS before sending a packet and appends the CS at the end of the packet and the R also computes CS over the packet data and

Fig. 3.67 Flowchart of
the oTx

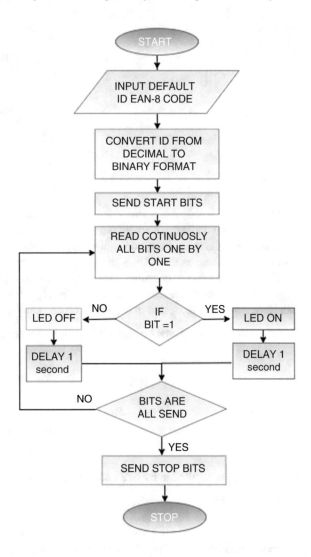

compares it to the CS appended to the transmitted packet. When the CS matched, the *R* sends an acknowledgment to the *T*. In case that CS doesn't match, means that errors during transmission occurred and no acknowledgment is sent. In this case, the *T* resends the packet.

However, there is a limitation of the CS. Since CS is a many-to-one mapping, there are several versions of the same string with the same CS, misleading the *R*.

CS algorithms are therefore designed to decrease as much as possible the probability of happening, not to eliminate it. When data (lamp ID) are stored, the EAN-8 code of each lamp is converted in binary as can be seen in the code written in C++ programming language presented in Appendix.

A logic representation of the algorithm, as flowchart, is presented in Fig. 3.67.

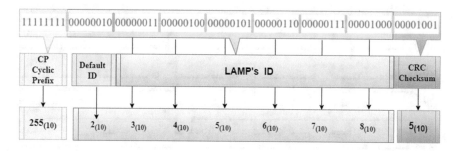

Fig. 3.68 Data stored into oTx driver with the VLC embedded. An example

The string of bits resulted and therefore data transmitted are (without spaces) the following:

$23{,}456{,}785_{(10)} = 00000010\ 00000011\ 00000100\ 00000101\ 00000110\ 00000111$
$00001000\ 00001001_{(2)}$

In order to send a packet that is easily recognizable by the oRx, a START string of bits (one byte) is additionally sent before data. The START string of 8 bits (one Byte) will be the cyclic prefix (CP) that is proposed in this book.

Since conversion from decimal to binary of any digital figures in EAN-8 code will have all the first four digits with value zero (from $i = 7$ to $i = 4$, where i—index) the START string of 8 bits will be:

$$CP = 255_{(10)} \leftrightarrow 11111111_{(2)}, \tag{3.38}$$

Therefore, the string ID of the lamp with the example above will be constantly sent as seen in Fig. 3.68.

Assigning one byte to the *ID default* and the last one to *CRC checksum*, the rest of 6 bytes are left to the lamp ID. So the total number of possible lamps and therefore valuable IDs for lamps calculate according to:

$$C_n^k = \frac{n!}{k!(n-k)!} \tag{3.39}$$

where:

n — 10 digits from 0 to 9
k — 6

$$C_{10}^6 = \frac{10!}{6!(10-6)!} = 210 \tag{3.40}$$

So, the total number of available lamp IDs in this case is 210.

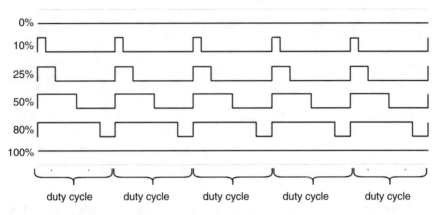

Fig. 3.69 Different duty cycles for the same frequency

This number can be increased up to 252 in case that only five digits (of the EAN 8 code) are considered for each lamp ID ($k = 5$), and the one left can have a different destination, for a more detailed description, as the company decides, therefore, finally 504 possible IDs can be used.

Encoding Data

A simple technique to send data by intensity modulation direct detection is using OOK, where the logic 1 corresponds to HIGH (LED is ON) and logic 0 corresponds to LOW (LED is OFF).

Encoding data with Manchester are one of the easiest and most reliable ways to send data since 1 is encoded as 10 and 0 as 01. However, the main drawback of this encoding technique is that long sequence of ones or zeros would result in errors of communication.

The intensity modulation direct detection is used for the VLC system with the OOK modulation and the signal encoding is done through PWM with a digital square wave. The frequency is constant and the duty cycle (the fraction of the time when the signal is on) can vary from 0% to 100% as it can be seen in Fig. 3.69.

PWM with Arduino can be done using de function:

```
analogWrite(pin, dutyCycle)
```

where:

1. Arduino board pin for PWM can be any of 3, 5, 6, 9, 10, or 11.
2. *dutyCycle*—a value from 0 to 255.

Fig. 3.70 Flowchart of the
oRx driver embedded in AP

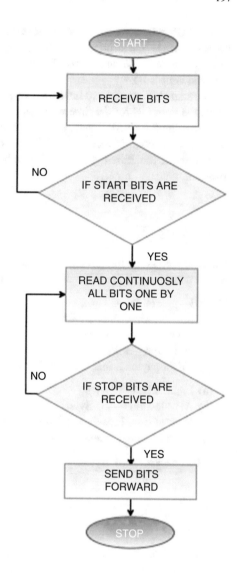

The main advantage of this method stands in the possibility of using any of the digital out pins of Arduino having, at the same time, under control both de duty cycle and the frequency.

On the other hand, as a drawback, any disconnection affects timings resulting in signal jitter. Codes written for wireless data communication of the VLC system are presented in Appendix A.

The oRx driver has to be carefully designed to make a proper trade-off between the distance between oTx and oRx for a proper communication and Bit Error Ratio (BER).

A logic representation of the algorithm, as flowchart, is presented in Fig. 3.70.

The AP with the oRx driver incorporated will receive, when in LoS and the proper distance, the string of bits sent by the oTx driver.

First, the start bits are received, being followed by the string of bits associated with the lamp's ID.

The starts bits considered here are a string of 1 Byte of ones. That means the oTx keeps the light ON for the time necessary to send 8 bits of ones.

Following the START byte, the oRx reads continuously the bits received till the entire EAN-8 code is received, sends the code forward to be checked, and then embedded in the Ethernet frame II together with the AP's ID (according to its location), date and time.

Ethernet Type II

The next step is the integration of the data received by oRx into an Ethernet frame type II consisting of the following bytes as in Fig. 3.71:

1. PRE—preamble—2 Bytes.
2. Destination MAC address—6 Bytes.
3. Source MAC address—6 Bytes.
4. Ethernet type—2 Bytes.
5. VLC data payload, that consists of:

   (a) Lamp's ID—8 Bytes.
   (b) AP's location—20 Bytes.
   (c) Date of signal received ID—10 Bytes.
   (d) Time of signal received ID—10 Bytes.

6. CRC checksum—4 Bytes.

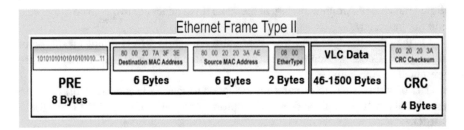

Fig. 3.71 Ethernet frame embedded in AP

Preamble—8 Bytes

The first string of bits in Ethernet frame with a predefined lengths and sequence aims to notify the controller that a packet is about to be sent, therefore, the controller is looking continuously while not already receiving a packet.

The oRx AP uses a preamble into the Ethernet frame that consists of two square waves with different period in sequence as follows:

```
01010101 01010101
```

Source MAC Address

Source MAC address consists, also of 48 bits in order to identify the specific AP underground.

Destination MAC Address

Destination MAC address is a unique string of 48 bits corresponding to the controller hardware in order to be easy identified by sender.

VLC Data

VLC data are the actual data that are going to be sent at the surface by each AP. This data are handed to the controller (its size is 48 Bytes but can be up to 12,000 bits).

Data consist of information received from the oTx (EAN-8 code aka miner's cap lamp ID), and those added by the AP itself (its own ID describing its location on the main gallery, time, and date of the received information from the oTx).

3.4 Integration and Testing

Tests in the laboratory have been done using a prototype of a main gallery profile as it is constructed underground precisely as its shape and reduced dimensions in the main galleries in Jiu Valley mines, with the same standardized shape. Dimensions of the prototype gallery are 1:6.25 according to original dimensions and shape.

The inside walls have been also painted (irregular black) in order to simulate the walls underground color black and gray covered with dust and tiny particles of rock (see Figs. 3.72 and 3.73).

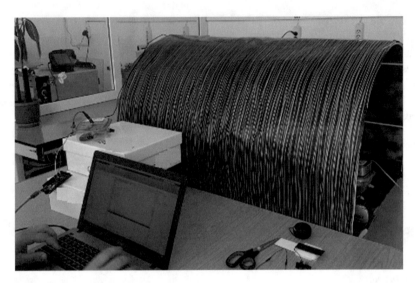

Fig. 3.72 Data acquired are stored on laptop

Fig. 3.73 LoS topology
setup inside the gallery
prototype (dimensions in
mm)

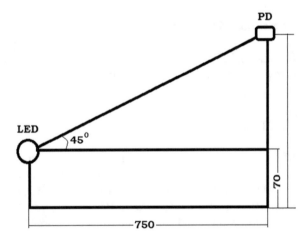

The LED and PD have been positioned with 45° tilt in order to have an FoV and
LoS topology as it is when the worker moves on the main gallery underground.

The oRx module (board with microcontrollers) is connected to the laptop. Data
acquired during communication are displayed on laptop and stored on the system
memory (Fig. 3.72).

In the prototype of the main gallery (that is with VLC prototype designed, the
distance between LED and PD is 750 mm (x) and the distance between LED and
bottom is 70 mm (y) (see Fig. 3.73).

The entire oTx is positioned into the prototype of the main gallery underground
(see Fig. 3.74.), the PIN PD (with regular optics) is positioned at the top of the

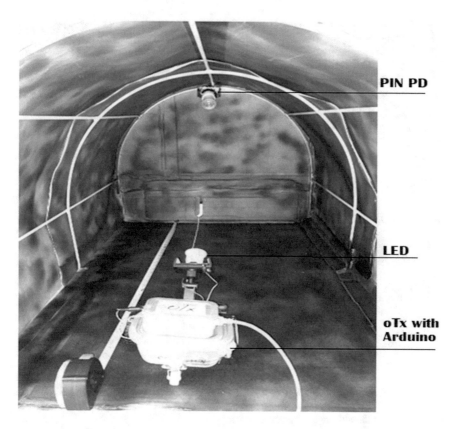

Fig. 3.74 The VLC setup inside the prototype of the gallery. Front view

gallery (see Fig. 3.75.) and the electrical setup with microcontroller of the oRx is placed outside the gallery in order to make it possible to test the system in a polluted environment (Fig. 3.76).

The UPO2104CS oscilloscope has been used to display data sent from the oTx with Arduino Uno and Lenovo Laptop (see Figs. 3.72 and 3.77).

In Fig. 3.77, the application Arduino is displayed with code written for sending data (the EAN-8 barcode embedded in the miner's cap-lamp) and the sent signal's shape is also seen on the display.

Figure 3.78 proves the signal received (the string of bits according to the lamp ID sent) at oTx on PD pins, without electronic circuit of TIA and microcontroller.

The low-cost VLC system designed and tested with a single optical Lambertian source (1 W cold white LED), without any optics in front of LED (biconvex lens), without a non-imaging concentrator or an optical band-pass filter in front of PD, the communication distance (tested into the main gallery prototype) reached 400 mm at a 4 kHz data rate.

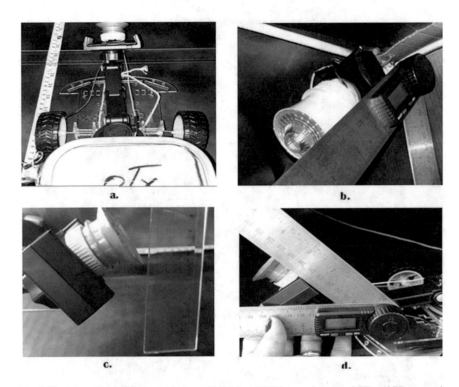

Fig. 3.75 Setups for simulation with the real situation. (1) Distance between LED and PD on x axis is 1000 mm; (2) PD tilt with angle 45 degrees on y axys; (3) c. LED positioned at 70 mm high from bottom; d. LED tilt with angle 45 degrees on y axis

The moving oTx module with LED (see Fig. 3.79) sends data by the optical signal to the fixed oRx module with a PIN PD. The equipment used during tests to generate, display, and acquire data consists of a function generator type MFG–8216A and an oscilloscope type UPO2104CS.

At the oTx side, the input electrical signal (channel 1/blue) is acquired with the oscilloscope connectors from the LEDs' terminals and at the oRx side (channel 2/yellow), the connectors are placed, first on the PIN PD terminals and then at the output of the TIA.

As it can be seen on the oscilloscope's display, the two signals (sent by LED embedded into the oTx and received by the PIN PD embedded into the oRx of the VLC system) are with different amplitudes but in phase. The output signal decreases in amplitude with the increased distance.

Data acquired are analyzed in order to determine the oRx' optimal key characteristics necessary for then AP positioned close to the working space underground.

An adaptive system has to be developed to keep working the VLC local wireless transmission even in proximity of the working face where the polluted environment will decrease the optical energy sent by LED.

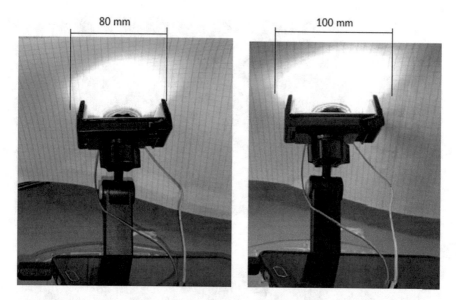

Fig. 3.76 Light beam spread in front of LED (without lens) 10 mm from LED (left) and 20 mm in front of LED (right). The optical distribution of LED used is checked at 10 and 20 mm in front of LED

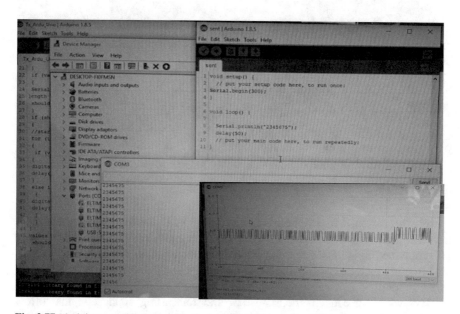

Fig. 3.77 Arduino app. Display with code data (EAN 8) and signal sent

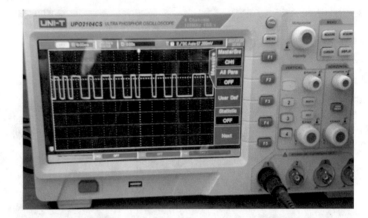

Fig. 3.78 Data (Lamp ID 23456785) received at PD's terminals seen on oscilloscope display

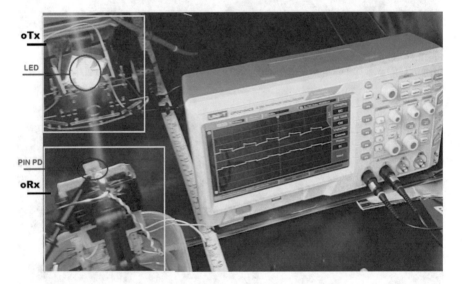

Fig. 3.79 Setup inside the gallery to complete tests

For this reason, data have been acquired during VLC data transmission and stored as *.csv files on the stick inserted into the USB port of the oscilloscope. All the *.csv files obtained during many tests have been then converted in order to be used as base information for an extended simulation and analyses done with the Eviews 10 software support.

EViews is an econometric, statistics, and forecasting application that offers powerful analytical tools within a flexible interface. It allows an efficient data managing, econometric, and statistical analysis. It also generates forecasts or simulates models, and produces high-quality graphs and tables for publication or inclusion in other applications.

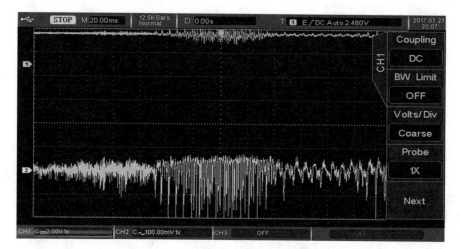

Fig. 3.80 Signals sent (channel 1 blue/up) and received (channel 2—yellow/down) displayed on the oscilloscope screen

They were used the sine wave signals to identify the system. This method has a number of advantages since it allows the direct determination of the frequency response of the system, ensures uniform accuracy over the whole frequency band of the studied system, and the internal noise is easily assimilated (Fig. 3.80).

Data acquired and analyzed are presented in Table 3.13.

As part of the system identification, the least squares method (LSM) is used to define the deterministic part model of a disrupted system using the mean square modeling error [34] (Fig. 3.81).

The model autoregressive controlled or with exogenous values (ARX) that are considered here, is defined by the following equations:

$$A(q) \cdot y(t) = B(q) \cdot u(t - k) + e(t) \tag{3.41}$$

where:

$$A(q) = 1 + a_1 \cdot q^{-1} + a_2 \cdot q^{-2} + \ldots + a_{na} \cdot q^{-na} \tag{3.42}$$

$$B(q) = b_1 \cdot q^{-1} + ab_2 \cdot q^{-2} + \ldots + b_{na} \cdot q^{-na} \tag{3.43}$$

where:

k—the dead time expressed in a number of sampling periods

$e(t)$—the prediction error

t—the normalized time (real time divided by the sampling period), the values from the set of integers

$u(t)$ the Input value at the time t

Table 3.13 Data analyzed with EViews software

Dependent variable: YA				
Method: Least squares				
Date: 04/20/20 time: 12:01				
Sample (adjusted): 26,100				
Included observations: 75 after adjustments				
Variable	Coefficient	SE	t-statistic	Prob.
UA	0.409548	0.080228	5.104817	0.0000
UA(−1)	0.155415	0.110339	1.408518	0.1718
UA(−2)	−0.067525	0.113482	−0.595023	0.5574
UA(−3)	−0.358969	0.113949	−3.150255	0.0043
UA(−4)	0.098018	0.130212	0.752755	0.4589
UA(−5)	0.017442	0.127617	0.136675	0.8924
UA(−6)	0.059017	0.131389	0.449178	0.6573
UA(−7)	0.023159	0.132004	0.175443	0.8622
UA(−8)	0.046535	0.123530	0.376710	0.7097
UA(−9)	0.087375	0.120678	0.724033	0.4760
UA(−10)	−0.158562	0.120953	−1.310938	0.2023
UA(−11)	−0.094388	0.116117	−0.812866	0.4243
UA(−12)	0.014529	0.113338	0.128194	0.8991
UA(−13)	0.041022	0.112011	0.366233	0.7174
UA(−14)	0.189087	0.111766	1.691811	0.1036
UA(−15)	−0.217951	0.127073	−1.715167	0.0992
UA(−16)	0.130822	0.128326	1.019451	0.3182
UA(−17)	−0.077746	0.130366	−0.596367	0.5565
UA(−18)	−0.181451	0.119279	−1.521229	0.1413
UA(−19)	−0.083728	0.116857	−0.716500	0.4806
UA(−20)	0.031608	0.112567	0.280791	0.7813
UA(−21)	0.066835	0.108730	0.614692	0.5445
UA(−22)	−0.238185	0.108630	−2.192636	0.0383
UA(−23)	0.063933	0.107087	0.597021	0.5561
UA(−24)	−0.131968	0.102794	−1.283820	0.2115
UA(−25)	0.173111	0.096742	1.789415	0.0862
YA(−1)	0.191010	0.170834	1.118103	0.2746
YA(−2)	0.125182	0.160318	0.780840	0.4425
YA(−3)	0.515259	0.168918	3.050356	0.0055
YA(−4)	−0.443199	0.197946	−2.238984	0.0347
YA(−5)	0.026109	0.194786	0.134039	0.8945
YA(−6)	−0.404687	0.196726	−2.057109	0.0507
YA(−7)	0.098673	0.202161	0.488091	0.6299
YA(−8)	−0.068035	0.190939	−0.356321	0.7247
YA(−9)	0.084199	0.188856	0.445837	0.6597
YA(−10)	0.310659	0.188079	1.651743	0.1116
YA(−11)	−0.226928	0.188214	−1.205688	0.2397

(continued)

Table 3.13 (continued)

Variable	Coefficient	SE	t-statistic	Prob.
YA(−12)	−0.023842	0.166739	−0.142991	0.8875
YA(−13)	−0.043527	0.173909	−0.250286	0.8045
YA(−14)	−0.020433	0.175039	−0.116732	0.9080
YA(−15)	0.069437	0.170073	0.408277	0.6867
YA(−16)	−0.163556	0.172398	−0.948715	0.3522
YA(−17)	0.031781	0.174303	0.182334	0.8569
YA(−18)	0.393100	0.168653	2.330821	0.0285
YA(−19)	−0.052877	0.174632	−0.302790	0.7647
Variable	Coefficient	SE	t-statistic	Prob.
Variable	Coefficient	Std. error	t-statistic	Prob.
YA(−20)	−0.097136	0.167498	−0.579923	0.5674
YA(−21)	−0.187870	0.171863	−1.093138	0.2852
YA(−22)	0.174134	0.167252	1.041150	0.3082
YA(−23)	−0.032103	0.161902	−0.198286	0.8445
YA(−24)	0.308557	0.175433	1.758834	0.0914
YA(−25)	−0.507815	0.166249	−3.054537	0.0054
R^2	0.926315	Mean dependent var		−0.008320
Adjusted R^2	0.772806	S.D. dependent var		0.136124
S.E. of regression	0.064883	Akaike info criterion		−2.411893
Sum squared resid	0.101036	Schwarz criterion		−0.836001
Log likelihood	141.4460	Hannan–Quinn criter.		−1.782657
Durbin–Watson stat	2.117332			

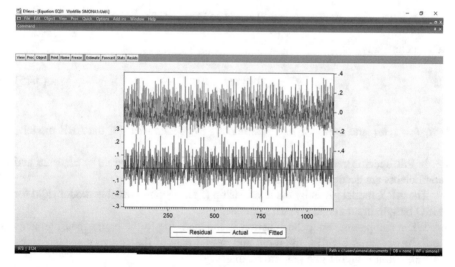

Fig. 3.81 Data processed with EViews support

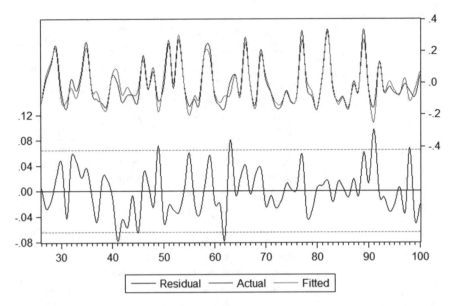

Fig. 3.82 Simulation results (red—real data acquired, green—estimated data, blue—error)

$y(t)$ the Output value at the time t
One-step delay operator

$$q^{-1} \cdot u(t) = u(t-1) \tag{3.44}$$

and accordingly:

$$q^{-k} \cdot u(t) = u(t-k) \tag{3.45}$$

$a_i; i = \overline{1, na}$ and $b_i; i = \overline{1, nb}$ parameters to be identified for the ARX model

In this specific case, the degree polynomials of A and B are $na = nb = 25$ and coefficients are according to Table 3.12.

The ARX model given by the relationship (5) is a single-variable model valid for SISO (single input single output) systems.

When the parameters determination is finished due to the entire measurements series processing, both for the input and output data of the system considered, the identification is an off-line parametric one.

The simulation results are presented in Fig. 3.82.

Up till now, due to a complex multipath CIR difficult evaluation and estimation, there are just a limited number of theoretical research in VLC technology applied in polluted industrial environments [35–40].

In this work, we take into consideration both intrinsic optical properties (IOPs) and apparent optical properties (AOPs) in industrial environments. The VLC setup complexity industrial environments increase with the polluted air since the optical path suffers from high loss because of both scattering and absorption optical properties. Since all these data are not persistent during a certain period of time and cannot be properly estimated, a VLC prototype has been developed and tests in laboratory have been done in order to acquire specific data both from oTx and oRx to simulate the entire VLC system.

According to data acquired and time series studied for both input and output data for the VLC system with LOS topology, based on *EViews* software, an ARX mathematical model results.

The determination coefficient (R2) is higher than 0.9 as can be seen in Table 3.12, therefore, a high-quality ARX mathematical model results.

This ARX mathematical model gives an accurate design regarding the VLC setup for an LoS topology in places with polluted environment, as the industrial facilities are.

The UP&MS based on the VLC system has the following advantages:

- allows the transmission of data identification through the visible light without affecting in any way the safety and/or personnel health in the potentially explosive underground environment,
- provide real-time information on the underground location of staff and visitors with the possibility of viewing on a digital map their position with a known margin of error in the underground lighting system dimensioning,
- allows the storage in a centralized database of all data relating to the position of the underground person at a given time; the database identifies the LED lamps based on the bar codes together with the date, time, and location at which they were identified,
- allows easy integration into any centralized system for staff data management,
- allows easy adaptation to other areas of activity such as other underground environments with more/less environmentally friendly conditions than those in the potentially explosive environment for which it is designed:

  1. is possible to be used for any type of underground lighting with fixed, networked lighting system,
  2. is possible to be used also in underground public transport systems (underground),
  3. is possible to be used also for underground mining spaces transformed into museums,
  4. is possible to be used in landscaped caves, integrated into the tourist circuit.

- each AP must periodically send (1–2 h) a state of its own status in order to check if the signal is correct. This feature identifies possible incidents, accidents, or malfunctions that are possible to occur underground.

References

1. [Online] https://www.cdc.gov/niosh/mining/content, Last accessed 9 August, 2020.
2. [Online] http://www.technowired.net/, Last accessed 9 August, 2020.
3. Carreño, J. P., Sousa e Silva, L., Almeida Neves, S. O., Aguayo, L., Braga, A. J., Noll Barreto, A., & Uzeda Garcia, L. G. (2016). Through-the-earth (TTE) Communications for Underground Mines. *The Journal of Computer Information Systems, 31*, 1.
4. [Online] http://www.gmggroup.org, White paper - Underground Mine Communications Infrastructure Guidelines – Part III: General Guidelines (2019) Last accessed 10 April, 2020.
5. Riurean, S., Olar, L. M., Leba, M., Ionica A. (2018). *Underground Positioning System Based on Visible Light Communication and Augmented Reality,* Modern Technologies for the 3rd Millennium 2018/3.
6. Riurean, S., Olar, M., Ionica, A., Pellegrini, L. (2019). *Using visible light communication and augmented reality for underground positioning system,* MATEC Web of Conferences SESAM2019, Vol. 305.
7. Rosca, S., Riurean, S., Leba, M., & Ionica, A. (2019). A reliable wireless communication system for hazardous environments. In T. Antipova & A. Rocha (Eds.), *Digital science. DSIC1, advances in intelligent systems and computing* (Vol. 850). Cham: Springer. (2018).
8. Wei, L., and Hagen, D. (2017). *Visible Light Communication. An alternative to the wireless transmission with RF spectrums through visible light communication,* University of Central Florida Department of Electrical Engineering and Computer Science EEL 4914 Design I 120 Pag. Sub.8/1/2017, Group 12 - CREOL Garrett Bennett, Benjamin Stuart, George Salinas and Zhitao Chen.
9. Zegong, L., Kicki, J., Xinzhu, H., Shujie, Y., Guanglong, D. and Sobczyk, E. J. (2017). *Mine Safety and Efficient Exploitation Facing Challenges of the 21st Century,* CRC Press, Published June 28, p. 300.
10. [Online] https://www.bongarde.com/niosh-conducting-illuminating-research-into-underground-mine-lighting/ Last accessed 10 April, 2020.
11. Sammarco, J. J., Reyes, M. A., & Gallagher, S. (2009). Do light-emitting diode cap lamps enable improvements in miner safety? NIOSHTIC2. *Mining and Engineering, 61*, 10. Number: 20036219.
12. Dickey, F. M., Holswade, S. C., & Shealy, D. L. (Eds.). (2005). *Laser beam shaping applications.* CRC Press.
13. Ma, H., Liu, Z., Jiang, P., Xiaojun, X., & Du, S. (2011). Improvement of Galilean refractive beam shaping system for accurately generating near-diffraction-limited flattop beam with arbitrary beam size. *Optics Express, 19*(14), 13105–13117. https://doi.org/10.1364/OE.19.013105.
14. Cannon, H. C. G. and George, W. (1943). *Refractive index of coals,* 09 Jan. 1943, Publisher Nature, Vol. 151.
15. Speight, J. G. (1994). *The chemistry and technology of coal, 1994.* New York: Marcel Decker.
16. [Online] https://phet.colorado.edu/sims/geometric-optics Last accessed 25 April, 2020.
17. [Online] https://my.zemax.com/en-US/Knowledge-Base, Last accessed 25 April, 2020.
18. [Online] http://everycircuit.com/app, Last accessed 10 April, 2019.
19. Pop, E., & Leba, M. C. (2003). *Microcontrollere si automate programabile.* București: Editura Didactica si Pedagogica.
20. Dimitrov, S., & Haas, H. (2015). *Principles of LED light communications. Towards networked Li-Fi.* Cambridge: Cambridge University Press.
21. Hulea, M., Ghassemlooy, Z., Abadi, M. M., Rajbhandari, S., & Tang, X. (2019). *Fog Mitigation Using SCM and Lens in FSO Communications.* 2019 2nd West Asian Colloquium on Optical Wireless Communications (WACOWC). doi:https://doi.org/10.1109/wacowc.2019.8770201.
22. Avatamanitei, S.-A., Cailean, A.-M., Zadobrischi, E., Done, A., Dimian, M., & Popa, V. (2019). *Intensive Testing of Infrastructure-to-Vehicle Visible Light Communications in*

Real Outdoor Scenario: Evaluation of a 50 meters link in Direct Sun Exposure. 2019 Global LIFI Congress (GLC). doi:https://doi.org/10.1109/glc.2019.8864129.

23. Ghassemlooy, Z., Popoola, W., & Rajbhandari, S. (2018). *Optical wireless communications system and channel modelling with Matlab* (2nd ed.). CRC Press.
24. [Online] https://www.gnu.org/software/octave, Last accessed 7 May, 2020.
25. [Online] https://www.engineeringtoolbox.com/refractive-index-d_1264.html, Last accessed 4 May, 2020.
26. Riurean, S., Leba, M., Ionica, A., Stoicuta, O. and Buioca, C. (2019). *Visible light wireless data communication in industrial environments*, IOP Conference Series: Materials Science and Engineering, Volume 572, International Conference on Innovative Research—ICIR EUROINVENT 2019 16–17 May 2019, Iasi, Romania.
27. Farahneh, H., Hussain, F., and Fernando, X. (2017). *A New Alarming System for an Underground Mining Environment Using Visible Light Communications*, IEEE Canada International Humanitarian Technology Conference (IHTC). doi:https://doi.org/10.1109/ihtc.2017.8058191.
28. Wei, L., Hagen, D. *Visible Light Communication. An alternative to the wireless transmission with RF spectrums through visible light communication*, University of Central Florida Department of Electrical Engineering and Computer Science EEL 4914 Design I 120 Pag. Sub.8/1/ 2017, Group 12 - CREOL Garrett Bennett, Benjamin Stuart, George Salinas and Zhitao Chen.
29. Vucic, J. et al. (2009). *125 Mbit/s over 5 m wireless distance by use of OOK-Modulated phosphorescent white LEDs*, 2009 35th European Conference on Optical Communication, Vienna, pp. 1–2.
30. Kang, W., & Hranilovic, S. (2008). Power reduction techniques for Multiplesubcarrier modulated diffuse wireless optical channels. *IEEE Transactions on Communications, 56*(2), 279–288.
31. Elgala, H., Mesleh, R., & Haas, H. (2011). Indoor optical wireless communication: Potential and state-of-the-art. *IEEE Communications Magazine, 49*(9), 56–62. https://doi.org/10.1109/ MCOM.2011.6011734.
32. Fuada, S., Putra, A. P., Aska, Y., & Adiono, T. (2016). *Trans-impedance amplifier (HA) design for Visible Light Communication (VLC) using commercially available OP-AMP.* 3rd International Conference on Information Technology, Computer, and Electrical Engineering (ICITACEE). doi:https://doi.org/10.1109/icitacee.2016.7892405.
33. Bhat, A. (2012). *Stabilize your transimpedance amplifier*, Application Note 5129, February.
34. Stoicuța, O., & Mândrescu, C. (2012). *Identificarea sistemelor. Lucrări de laborator.* Petrosani: Editura Universitas.
35. Wang, W.-Z., Wang, Y.-M., Shi, G.-Q., and Wang, D.-M. (2015). *Numerical study on infrared optical property of diffuse coal particles in mine fully mechanized working combined with CFD method*, Hindawi Publishing Corporation, Mathematical Problems in Engineering, Volume 2015, Article ID 501401, 10 pages doi:https://doi.org/10.1155/2015/501401.
36. Wang, J., Al-Kinani, A., Zhang, W. and Wang, C. (2017). *A new VLC channel model for underground mining environments*, In: 13th International Wireless Communications and Mobile Computing Conference (IWCMC), Valencia, pp. 2134–2139.
37. Okada, Y., Nakamura, A. M., & Mukai, T. (2006). Light scattering by particulate media of irregularly shaped particles: Laboratory measurements and numerical simulations. *Journal of Quantitative Spectroscopy & Radiative Transfer, 100*, 295–304.
38. Tang, H., & Lin, J.-Z. (2013). Retrieval of spheroid particle size distribution from spectral extinction data in the independent mode using PCA approach. *Journal of Quantitative Spectroscopy and Radiative Transfer, 115*, 78–92.
39. Mishchenko, M. I., Lacis, A. A., Carlson, B. E., and Travis, L. D. (1995). Nonsphericity of dustlike tropospheric aerosols: Implications for aerosol remote sensing and climate modeling, Geophysical Research Letter, Vol.22, Iss.9, Pag.1077–1080, doi: https://doi.org/10.1029/ 95GL00798.
40. Avery, R. K., & Jones, A. R. (2014). Measurement of the complex refractive index of pulverised coal by light scattering: An attempt and some comments. *Journal of Physics, 15*, 8. https://doi. org/10.1088/0022-3727/15/8/008.

Notations

$\otimes$	Convolution
A	Attenuation of optical signal [dB/m]
a	LED's semi diameter
$A_{\text{col PD}}$	Collection area of PD
A_j	The junction area
A_{PD}	Photodetector's active area
B	Bandwidth
b	PD's semi-diameter
	Semi-length of ray in lens
c	Speed of the light in free space $(3 * 10^8$ m/s)
$C_b\text{IR}$	channel's baseband impulse response
C_d	LED's capacitance
C_d	Sum of the junction capacitance
C_j	Distributed capacitance
d	Distance between oTx and oRx on the direct path
d	Semi-diameter (marginal ray) of lens
d_1	Distance between oTx and the reflecting object
d_1	Distance from LED to lens' centre (LEN)
d_2	Distance between the reflecting object and oRx
e	Distance from Ob to LED
e	Elementary electron charge $e = 1.602 \times 10^{-19}$ [C] (coulombs)
E	Photons' energy
E_{RMS}	RMS value of a signal $(x(t))$
f	Focal distance of lens
f	Frequency corresponding to the photon wavelength
f_c	Carrier's frequency
f_{cLED}	LED's cut-off frequency
f_n	The frequency of the nth subcarrier
FoV	Field of View

© Springer Nature Switzerland AG 2021
S. M. Riurean et al., *Application of Visible Light Wireless Communication in Underground Mine*, https://doi.org/10.1007/978-3-030-61408-9

f_s	Sampling rate
g	Distance from PD to Im
$g(\omega)$	Optical gain of an ideal non-imaging concentrator
G_{APD}	Gain of the APD PD
G_{oc}	Optical concentrator's gain
G_{TIA}	TIA's gain
h	Planck's constant (6.626×10^{-34} Js)
h	Planck's constant (6.62×10^{-34} J or $4.135\ 10^{-16}$ eV s/rad)
$h(t)$	Optical CIR
$H_{\text{CIR}}(f)$	Channel's frequency response
$H_{\text{D/A}}$	Digital to analog converter response
$H_{\text{DAC}}(f)$	The frequency response of D/A converter
H_{dif}	The transfer function in environments with diffused link
h_{FE}	DC current gain
H_{LoS}	DC gain of oRx
$H_{\text{VLC}}(f)$	Transfer function for VLC transmission
$H_{\text{LED}}(f)$	LED's frequency response
H_{LoS}	The transfer function in Line of Sight (LoS) scenario
I_d	Dark current
I_F	Forward biased LED current
I_G	Value of the output current, amplified due to avalanche effect
I_p	Current without amplification
I_{PD}	Generated photocurrent by PD
k	The Boltzmann's constant (1.38×10^{-23} [J/K]) [146]
$m(t)$	Unipolar binary message signal that has to be transmitted
m_l	Lambert's mode number expressing directivity of the oTx beam related to the LED's semi-angle at half-power $\varphi_{1/2}$
n	Internal refractive index
N	Number of subbcariers used
n	The refractive index of the concentrator
$n(t)$	Total noise (ambient noise in optical channel, shot noise and thermal noise)
n_{air}	Refractive index of air
N_{CP}	Length of cycle prefix CP
n_{glass}	Refractive index of glass
n_s	Shot noise
n_t	Thermal noise
N_u	Number of subcarriers used to carry data
oRx	Optical receiver
oRx	Optical transmitter
$P(\varphi)$	Radiant power intensity
P_c	Clipping probability
$\overline{P}$	Average optical power
$P_{\text{LoS\_dp}}$	Optical power received from LoS direct path topology

$P_{\text{NLoS\_sr}}$	Optical power received from a single reflection coming from the reflecting surface
P_{o}	Optical power emitted by a light source
$P_{\text{PD\_LOS}}$	Optical intensity of signal received by PD in LoS scenario
$P_{\text{sd}}(\lambda)$	Radiation power spectrum of the LED
$Q(r_i)$	Q-function for the tail probability of a Gaussian distribution
R	Radius of curvature (biconvex lens)
$\mathcal{R}$	The photodetector's responsivity
RMS n_{d}	Root-mean-square of dark current noise
RMS n_{t}	Root-mean-square of thermal noise
$R(\theta_i)$	Bidirectional reflectance distribution function
r_1 and r_2	Clipping ratios
$\mathcal{R}_{\text{APD}}$	The PD APD's spectral responsivity
R_{b}	Bit rate
R_{c}	Sum of the series resistance and the load resistance of the PD
r_{os}	Oversampling ratio
$\mathcal{R}_{\text{PIN}}$	The PD PIN's spectral responsivity
R_{r}	Parallel result of the load resistor and the amplifier input resistor
R_{s}	OFDM symbol rate
R_{s}	Series resistor
$R_{\text{SD(on)}}$	Total resistance in the path from source to drain
S	Optical path length
S	Photosensitivity
$s(t)$	The ASK output signal
sr	Steradian
T	Equivalent noise temperature
t	Time
t_0	The minimum time delay
t_{d}	Delay time
t_{f}	Fall time
t_{r}	Rise time
$T_{\text{s}}(\omega)$	Optical transmission of the band-pass filter
$V(\lambda)$	Relative eye sensitivity (is normalized to unity at the peak wavelength of 555 nm)
V_{BE}	Base-Emitter Saturation Voltage
V_{CE}	Collector-Emitter Saturation Voltage
V_{F}	Forward voltage drop
w	Thickness of the I-layer
$x(t)$	Transmitted optical signal
$x_{\text{c}}[k]$	Clipped signal
X_n	Symbol sent at the nth subcarrier
$y(t)$	Signal received by photodetector
$\alpha(\lambda)$	The absorption coefficient
$\delta(.)$	Dirac function

$\delta(t - d/c)$	Signal propagation delay
ε	Dielectric constant
η	Luminous/quantum efficiency
θ_i	Incident angle of the outgoing light
θ_o	Observation angle of the incoming light
λ	Wavelength of the absorbed/emitted light
μ	Mean delay spread (DS)
ρ	Reflexion coefficient of the reflecting object according to its material's surface
σ	The variance of the signal samples distribution
σ_{AWGN}	Electrical gain
σ_τ	Signal's temporal dispersion
τ_i	Signal's excess delay
φ	Half-angle FoV
$\varphi_{FoV,oTx.}$	Radiation pattern FoV of oTx
Φ_V	Luminous flux
ω	Angular frequency
ω	Radiation incident at angle
$\omega_c \leq \pi/2$	FoV
φ	Angle of irradiance with respect to the axis normal to the transmitter surface for maximum radiated power

Appendix

Program that calculates the checksum digit for EAN 8

```
public static String generateEAN(String barcode) {
    int first = 0;
    int second = 0;

    if(barcode.length() == 7 || barcode.length() == 12) {

        for (int counter = 0; counter < barcode.length() - 1; counter++) {
        first = (first + Integer.valueOf(barcode.substring(counter, counter
+ 1)));
            counter++;
            second = (second + Integer.valueOf(barcode.substring(counter,
counter + 1)));
        }
        second = second * 3;
        int total = second + first;
        int roundedNum = Math.round((total + 9) / 10 * 10);

        barcode = barcode + String.valueOf(roundedNum - total);
    }
    return barcode;
}
```

Program (written in C++ programming language) that converts decimal $_{(10)}$ into binary $_{(2)}$ - Version 1

```
// C++ program that converts decimal (10) into binary (2)
#include <iostream>
using namespace std;

// function for decimal to binary conversion
void DecToBin(int n)
```

© Springer Nature Switzerland AG 2021
S. M. Riurean et al., *Application of Visible Light Wireless Communication in Underground Mine*, https://doi.org/10.1007/978-3-030-61408-9

```cpp
{
// array for binary storage
int binaryNum[1000];

// counter for binary array
int i = 0;
while (n > 0)
{

// storing remainder in binary array
 binaryNum[i] = n % 2;
n = n / 2;
i++;
}

// binary array is printed in reverse order
for (int j = i - 1; j >= 0; j-)
cout << binaryNum[j];
}

// driver program used to test the function above
int main()
{
   int n = 7;
   DecToBin(n);
   return 0;
}
```

Program (written in C++ programming language) that converts decimal $_{(10)}$ into binary $_{(2)}$ - Version 2 - using the bitwise operator

```cpp
// // C++ program that converts decimal to binary using bitwise
operator
// Size of an integer is 8 bits
#include <iostream>
using namespace std;

// function for decimal to binary conversion
int DecToBin(int n)
{
   // Size of an integer is 8 bits
   for (int i = 7; i >= 0; i-) {
      int k = n >> i;
     if (k & 1)
       cout << "1";
     else
       cout << "0";
   }
}

// driver code
int main()
```

```
{
  int n = 8;
  DecToBin(n);
}
```

Code for ID communication:

```
void setup() {
  pinMode(5, OUTPUT);
}
void loop() {
  digitalWrite(5, HIGH);
  delayMicroseconds(100); // A 10% duty cycle is about 1KHz
  digitalWrite(5, LOW);
  delayMicroseconds(1000 - 100);
}
```

Code for sending digital 6 (for example), from oTx to oRx (version 1)

```
  Serial.print("Insert digit:");
  Serial.println(" ");
}

void loop() {
  digitalWrite(LEDTrans,HIGH); //turns the LED on
  delay(timeWaitOn); //delay in milliseconds (1 sec)
  digitalWrite(LEDTrans,LOW); //turns the LED off
  delay(timeWaitOff); //wait a second
    /*checks if data has been sent from the computer: */
  if (Serial.available()) {
    byteRead = Serial.read();
    //will print the binary representation of 'byteRead' as 8 characters
of '1's and '0's, Most Significant Bit first.
  for (uint8_t bitMask = 128; bitMask != 0; bitMask = bitMask >> 1) {
      if (byteRead & bitMask) {
    Serial.write('1');
    } else
  {
    Serial.write('0');
  }
  }
  Serial.write("  ");
  Serial.println(byteRead); //send back the actual ASCII code
  }
}
```

Code for sending digital 6 (for example), from oTx to oRx (version 2)

```
  const int LEDTrans = 13;
  const int timeWaitOn = 1000; // millisec (1 sec)
  const int timeWaitOff = 1000; // millisec (1 sec)

  void setup() {
    // Turn ON the Serial Protocol
    Serial.begin(9600);
    Serial.println("Send digit:");
    pinMode (LEDTrans, OUTPUT);
```

```
    }
  void loop() {
   /*check if data has been sent from the computer: */
   if (Serial.available()) {
    byte byteRead = Serial.read();
    // will print the binary representation of 'byteRead' as
8 characters of '1's and '0's, Most Significant Bit first.
     for (uint8_t bitMask = 128; bitMask != 0; bitMask >>= 1 ) {
      if (byteRead & bitMask) {
       digitalWrite(LEDTrans, HIGH); //turns the LED on
   delay(timeWaitOn); //delay in milliseconds as for example 1 second
       digitalWrite(LEDTrans, LOW); //turns the LED off
      } else {
       delay(timeWaitOff); //wait 1 second
      } // end of if
     } // end of for loop
    } // end of if something available
   } // end of loop
 SEND EAN 8
  void setup() {
  Serial.begin(300);
  }

  void loop() {

   Serial.println("23456785");
   delay(100);

 Receive EAN 8
  void setup()
  {
   Serial.begin(300);
  pinMode(8,OUTPUT);
  }
  void loop() {
   // digitalWrite(8,HIGH)
```

Index

A

Additive White Gaussian Noise (AWGN), 10, 48
Amplifier circuit, 81
Amplitude modulation (AM), 103
Amplitude shift keying (ASK), 101
Analog oTx drivers, 82
Analog to digital converter (ADC), 48
Analogue telephone adapter (ATA) devices, 130
Angle of arrival (AoA), 26
Arduino Mega board, 191
Arduino Mega's characteristics, 172
Arduino programs, 162
Arduino software, 172
Arduino Uno board, 185
Arduino Uno's main characteristics, 164
ARX mathematical model, 209
ARX model, 208
Attenuation, 147
Avalanche photodiode (APD), 69

B

BeamCaster project, 20
Biconvex lenses, 148
Bidirectional reflectance distribution function (BRDF), 53
Bipolar signals, 108
Bit Error Ratio (BER), 12, 197
Blu-ray discs, 58
BPW20R PIN PD, 169

C

Cable-based communication, 43
Ceiling bounce model, 88
Cellular networking, 130
Channel DC gain, 177
Channel impulse response (CIR), 10, 44, 45, 48, 96
Channel model, 86
 light distribution indoor, 174
 oTx driver setup, 173
Channel RMS DS, 92
Checksum (CS) algorithm, 193
Cisco Visual Networking Index (CVNI), 6
Coal mining environment, 98
Coal mining industry, 134
Coded Modulation (CM), 12
Colour Rendering Index (CRI), 15
Commulight, 19
Communication system, 132
Computer numerical simulations, 176
Constant Current Reduction (CCR), 13
CSK modulation technique, 113
Current-mode analog drivers, 82
Cycle prefix (CP), 105
Cyclic prefix (CP), 103

D

Darlington pair configuration, 156
Darlington transistor configuration, 156
Data communication, 98
DC reduced HCM (DCR-HCM), 112
DC–DC converter, 78

© Springer Nature Switzerland AG 2021
S. M. Riurean et al., *Application of Visible Light Wireless Communication in Underground Mine*, https://doi.org/10.1007/978-3-030-61408-9

Printed in the United States
by Baker & Taylor Publisher Services